On a Silver Platter

On a Silver Platter

CD-ROMs and the Promises of a New Technology

EDITED BY

Greg M. Smith

New York University Press
NEW YORK AND LONDON

NEW YORK UNIVERSITY PRESS
New York and London

Library of Congress Cataloging-in-Publication Data
On a silver platter : CD-ROMs and the promises of a new technology / edited by Greg M. Smith.
p. cm.
Includes bibliographical references and index.
ISBN 0-8147-8080-6 (cloth : acid-free paper)
ISBN 0-8147-8081-4 (paper : acid-free paper)
1. CD-ROMs. 2. Interactive multimedia. I. Smith, Greg M., 1962–
TK7895.C39 05 1999 006.7—ddc21 98-25494
CIP

New York University Press books are printed on acid-free paper, and their binding materials are chosen for strength and durability.

Manufactured in the United States of America

10 9 8 7 6 5 4 3 2 1

Contents

Chapter One

Introduction
A Few Words about Interactivity

Greg M. Smith

The Seduction of the Potential

Multimedia, hypermedia, and CD-ROMs all offer the potential to reshape storytelling—or so we are told. New media will someday allow us to navigate in a virtual story world, to choose what events we are interested in, and to alter the course of those events. Such new texts will transform what might be called the "reader" or "spectator" into an "interactive player" who controls the flow and direction of the text. Instead of being passive recipients of stories, we will be cocreators and participants in the narratives of the future, which will allow us to "travel" to new and compelling virtual worlds full of choice. "You ain't seen nothing yet," we are promised.

Critics writing about new technology have tended to emphasize the vast "potential" of hypermedia and multimedia. All too often, however, the realities of actual CD-ROMs fall short of the utopian visions promised by such writing. A CD-ROM praised for its "virtual reality" can be difficult to install, slow to run, and frustrating to navigate. If we confuse the promises of a new medium with the pitfalls and payoffs of actual texts, it becomes difficult to see exactly what a new medium (such as the CD-ROM) really does.

Their technological drawbacks notwithstanding, CD-ROMs today provide fascinating experiences. Without being grounded in actual texts and/or specific reception contexts, however, CD-ROM criticism can easily miss what is fascinating about *these particular CD-ROMs*. Without being anchored in specific examples, critics can fall into the trap of echoing the self-promoting rhetoric of CD-ROM developers.

On a Silver Platter asserts that multimedia and CD-ROMs are *actual* media and are no longer merely *potential* media. We need no longer spend the bulk of our time speculating on what forms these media will take, since interesting examples now exist. We should recognize that there are complex CD-ROMs out there that are just as worthy of analysis as films or literary texts. We need to acknowledge that these multimedia texts are being integrated into real people's lives in ways that deserve to be studied. This anthology is intended to announce a kind of "coming of age" of CD-ROMs as a commercially, socially, and aesthetically significant medium worthy of close critical attention by media scholars.

When academics do examine particular new media texts and the contexts of their reception, they frequently examine somewhat avant-garde texts that seem to hold the keys to the radical potential of new media. For instance, Michael Joyce's hypertext *Afternoon, a story* has received more scholarly attention than the blockbuster CD-ROM *Doom,* although only a fraction of new media users have heard of Joyce's innovative text.[1] By scrutinizing relatively noncommercial multimedia art texts, academics continue to emphasize that these new media are more interesting for their potential capacities. To this way of thinking, avant-garde texts are more likely to give us a glimpse of the medium's future promise than more commercial products would.

Scholars have also given significant attention to the online world of the Internet, particularly the interactivity of chat rooms, IRCs (Internet relay chats), and MUDs (multiple user dungeons/domains). Sherry Turkle's (1995) and Allucquere Rosanne Stone's (1995) fascinating ethnographic inquiries into the social world of the Internet have called into question basic concepts of personal identity. In a chat room we can explore what it might be like to occupy another subjectivity by creating a range of alternate electronic selves, destabilizing the notion of a unified personality. Such interactions allow us to ponder the meaning of race, gender, and sexual orientation in a virtual environment. Such experiences offer the promise of at least temporarily escaping the "meat" of one's own body (to use William Gibson's term from *Neuromancer*) in favor of a less constraining realm of freedom. Turkle's and Stone's work reports (and, at times, criticizes) chat room participants' optimism about the possibility of radically destabilizing age-old gender and race categories. Although Stone and Turkle at times criticize and qualify this optimism, they are obviously drawn to the subversive potential this new realm presents.

What seems to be neglected between the avant-garde and the online is the multimedia form that has come to be the most commercially successful: the CD-ROM. *Quake* or *Wing Commander* may not offer the radical restructuring of narrative form that *Afternoon* does or the possibility for identity-reconceptualizing masquerade that online chat rooms do. However, this relative lack of subversive potential does not make them insignificant. In fact, CD-ROMs exert a powerful influence on the way we think about new media.

Our understanding of multimedia is shaped by exaggerated hype, of course, but it is also formed during our interactions with actual CD-ROMs. We come to understand what "interactivity" is through our experiences with these texts. The widespread use of CD-ROMs, at a time when new media are being defined, makes them significant.

On a Silver Platter examines particular CD-ROM texts and the contexts of their reception. *Particularity* is the watchword for this collection. Through this particularity, the authors hope to complicate the utopian discourses about new technologies, and make more grounded assertions about the present and future of the medium. I argue that close attention to encounters with such early texts can yield invaluable hints of what is to come. Since the future is an outgrowth of the present, particular attention to present multimedia can give much more insight into the future than broad prognostications about the medium's potential.

My hope is that by engaging in criticism of current multimedia texts, media scholars can help define the future of these new media. Academic media studies tends to wait until long after a medium has become widely accepted before it gives the medium close critical attention (for example, film studies emerged in the 1960s and television studies in the 1980s). By neglecting to provide detailed criticism of new media texts, scholars pass up the opportunity to shape the future of media.[2]

Just as encounters with popular texts can shape the future of a medium whose conventions are still being defined, so can criticism articulate understandings that are only implicitly voiced in the commercial sphere. Our definition of multimedia is shaped by publicity discourses and actual interactions with CD-ROMs, but it can also be affected by criticism that opens up new ways of thinking about the medium. Media scholars should learn the lessons of their past and intervene early in the development and reception of multimedia by doing what they do best: analyzing texts and contexts with specificity and insight. *On a Silver Platter* presents scholarship that does exactly that.

The Betamax of the Nineties

A book about particular CD-ROMs runs the risk of becoming obsolete when the CD-ROMs are no longer in wide use. In fact, some have argued that CD-ROMs themselves are an endangered species. Although CD-ROM drives are now standard equipment on most new computers and CD-ROMs are the dominant multimedia format in sales, there are factors that seem to presage the death of the silver disc. Although the early advantage of the CD-ROM was its ability to store a seemingly enormous amount of data, the individual CD-ROM's memory capacity is now becoming too limited as the complexity of programs increases.

Already it feels like a single CD-ROM is not enough. As I eject and substitute individual discs while playing a five CD-ROM game, I am reminded of turning vinyl records over on my now-defunct turntable, and I can feel the imminent obsolescence of the new medium. The greater storage capacities of the DVD (digital video disc) may eventually allow it to replace the CD-ROM as the medium of choice. Or perhaps Internet CD-ROMs will gain superiority by providing the user with a basic platform, which then points him/her to the Net to access the full database (which the manufacturer can update without issuing new CD-ROMs).[3]

What if CD-ROMs do become obsolete? Even then, I believe that understanding them is crucial to understanding multimedia, because early attempts to work in a new medium have long-lasting effects on the later history of that medium. Historians investigating the early practice of cinema have discovered how these early films shaped the norms of the classical cinema. As early film practitioners experimented in making texts, they slowly built a set of conventions for how to shoot and edit a scene, and these conventions were internalized by audiences and filmmakers alike. The fundamental expectations for what a film is and how it tells its story were formed before 1917.

Although these pre-1917 films may seem outdated to modern audiences, they were important in shaping the subsequent development of the medium. Similarly, even if CD-ROMs become relics, they will still exert an influence on later interactive media, because it was with CD-ROMs that many of us got our first taste of computer "interactivity." During these early interactions with CD-ROMs, we are developing conventions for what interactivity is, and these conventions will undoubtedly shape the future of the medium.

A medium's potential does not exist somewhere in space waiting to be discovered. It is created through the practice of making marketable texts for a mass audience. Although the future of new media may seem to lie in more radically innovative texts, modern media history tells us that the broad understanding of a medium is formed through early widespread experiences of commercialized texts. Perhaps more than anywhere else, we learn what this new concept of interactivity is by playing/using actual CD-ROMs, and in new media, first impressions have a lasting bearing on later expectations.

Those who suggest that CD-ROMs are a relatively unimportant temporary solution while we wait for high-speed Internet access are making a fundamental error. They confuse the importance of delivery systems with the importance of development systems. Development systems are the tools used by software designers to create interactive texts. They include graphic rendering programs, animation software, sound design systems, and multimedia authoring programs such as Macromedia Director. Delivery systems, on the other hand, are the tools for getting the finished text to the consumer. Internet Web access is one such system, and CD-ROMs are another.

Even if future texts are no longer delivered on CD-ROM, the process of designing such interactive texts will strongly resemble the process of designing a CD-ROM. No one believes that HTML, the simple language for creating World Wide Web pages, is a sufficiently subtle format for creating intricate interactive texts such as *Myst*. Even if *Myst*'s descendants are delivered to customers via high-speed Web connections, creating these texts will be more like designing a CD-ROM than designing a Web page. Designing CD-ROMs involves much more complex and expressive capabilities than are possible with HTML or Java, so the CD-ROM will continue to be the model for designing complex interactive texts, regardless of how those texts are delivered. Even if CD-ROMs as delivery systems decline, they will continue to exert their influence for the foreseeable future as exemplars of complicated, nuanced interactive texts.

Designed and Bounded

The subject of this book, then, is defined less by the actual medium in question (the CD-ROM) than by the nature of the experiences the medium offers. Even if CD-ROMs as a physical medium vanish in the not-too-distant future, players/users will continue to play *Sim City*, *Command and Conquer*, or their descendants. How do we talk about the experiences CD-ROMs offer without linking them to a particular physical medium? If we will no longer play CD-ROMs, what will we be playing?

The experiences currently provided by CD-ROMs are quite different from the interactivity provided by online chat. First, the CD-ROM presents itself as a strongly designed object. People interact with *Doom* in order to experience the particular blend of violence and mayhem designed into the *Doom* universe. Like a film or television program, the CD-ROM has been authored and constructed to provide different audience members with similar experiences. Such CD-ROMs are designed by a few people to be experienced by many, and these few-to-many media bear the imprints of their creators. Although they are sometimes described as "virtual reality," they are much more highly shaped by human intent than any real world. They have been authored to make the fictional world more interesting; the dull parts of reality have been left out.[4] Such CD-ROMs are compelling not just because they present a realistically cohesive world but because that world has been shaped and designed to provide significant payoffs for our time investment.

Of course Internet chat rooms are purposefully designed too. They are established with the metaphor of four interior walls, and they promote a particular kind of interaction (simulated conversation through real-time typing). However, they are not *strongly* designed. People do not "visit" a chat room to experience a cleverly designed universe. They use chat rooms as neutral spaces in which to "meet" other people interested in similar topics. The appeal of a chat room is not in the "room" itself but in the discussions one has with people who share your interests. The room itself is not compelling, although the conversations one has in it may be.

The difference between a strongly designed object (like a CD-ROM) and a weakly designed object (like a chat room) is like the difference between a furnished house and an unfurnished one. When entering a furnished house, one reacts to the design and decorating of the space, which bear the imprint of the owner's tastes. The furnishings convey what activities are expected in those rooms; clearly the bedroom is for sleeping and the dining room is for eating. When one enters an unfurnished house, one gets more of a sense of the possibilities of the space. A particular room might be used either as a bedroom or a dining room, and the tone of the house is only broadly defined by the architectural design. The pleasures of an unfurnished room lie in its possibilities; the pleasures of a furnished room reside in the sense of an already actualized design concept.

The furor over "interactivity" has often emphasized the "design-your-own" possibilities of future media. What fun it will be to create our own stories, bend narratives to our will, wield a creator's power over a virtual world! And yet it is hard to imagine such do-it-yourself narration ever entirely supplanting the experience of a well-authored world, primarily because it's just damn hard to create a compelling world. We go to strongly designed texts like CD-ROMs and movies and television programs because we want to experience a story/world that has been well-made for our consumption. We may enjoy holding up our end of an online conversation, but that also requires considerable mental and social effort. A large part of the lure of CD-ROM texts is the promise of experiencing a virtual world that is more interesting than we probably could design for ourselves. Just as inexpensive home video equipment did not destroy the Hollywood video market, just as the relative ease of writing on a word processor did not create a surge of new novelists, the ability to "roll your own" texts will not supplant the desire to experience a text that has already been well-made for you. In spite of the growth of other kinds of interactive experiences, there will always be a market for the strongly designed medium.[5]

CD-ROMs share with films and novels this sense of being strongly designed. However, they significantly differ from those other media, which attempt to prescribe a sequence of narrative events. The distinction between the two is not a strong dichotomy between the older "linear" and the newer "nonlinear" media, as I will argue later in this introduction. Rather, I suggest that these media differ in the boundaries they establish for the player's/viewer's/reader's interactions with the strongly designed medium.

I would like to describe "interactivity," as it applies to strongly designed computer media such as CD-ROMs, in terms of boundaries. In many (but not all) cases, multimedia offers an interactivity that is less strongly bounded than that of books, films, videos, and so forth. The difference is not that one medium is interactive while the other is not. A more fruitful way of thinking is that interactivity among strongly designed media differs according to certain parameters. The next sections will focus on the parameters that define the experience of CD-ROM interactivity.

To talk about "boundaries" may seem like a strange way to discuss multimedia, since the hype over multimedia rarely mentions the constraints of the new medium. But according to theorist Rudolf Arnheim (1957), the boundaries and limits of a medium are as crucial to its definition as the medium's potentials.

In film, for instance, the image cannot exceed a stable boundary: the size of the screen. This certainly is a limitation on what the film medium can and cannot do. For Arnheim, however, such limitations created a new parameter for aesthetic expression: offscreen space. Since the medium necessarily distinguishes between offscreen and onscreen space, the filmmaker can play with this parameter in aesthetically powerful ways. Horror filmmakers, for instance, can toy with our expectations of when the monster (lurking just outside the onscreen space) will appear. Arnheim emphasized that the potential for film to play this game of cat-and-mouse is made possible by the limitations of the medium.

Arnheim, however, did not emphasize the way a medium evolves through actual practice. He generally treated the film medium as a Platonic ideal whose potentials were waiting to be uncovered. While I wish to follow Arnheim's instinct in showing the relationship between a medium's boundaries and its capacities, I do not wish to duplicate his idealism of media characteristics. A young medium, as I asserted earlier, develops its "unique" properties as developers create new texts and readers interact with them. This means that the definitions of multimedia or hypermedia or strongly defined objects are not based solely on their aesthetic potentials. Instead, our definitions of new media are forged largely in the commercial sphere, where economic factors play an enormous part.

The form of interactivity that exists today is shaped by the desire to create CD-ROMs that encourage consumers to buy them, play them, and then buy more. This simple fact has a strong (though often ignored) effect on the kind of interactivity we experience in playing current CD-ROMs. Our understanding of interactivity is shaped as much by commercial forces as it is by the hypothetical aesthetic potential of the medium itself. The evolving definition of interactivity is shaped by social, economic, and aesthetic forces, and the following discussion seeks to tease out these forces so we can examine them more closely.

The Fantasy of Interactivity: Active Choice and Control

The word "interactive" is used so frequently to publicize CD-ROM products that it has become a defining characteristic of the medium. Yet this defining characteristic seems to have no central definition. "Interactivity" seems like the Supreme Court's description of pornography: I know it when I see it. It is both a goal to which CD-ROMs aspire (and thus a description of the medium's future) and a description of the present state of the medium. This term gains meaning as it is used and circulated. As more and different CD-ROMs appear on the market, they influence our understanding of just what interactivity is. It is a term defined by practice: through the interchanges among players buying and using CD-ROMs, mass media writers describing them, and designers creating them.

One project for the scholarly study of CD-ROMs is to propose potentially useful ways to conceptualize interactivity. The rest of this introduction attempts to enumerate some of the things that are meant by the word "interactive" as it applies to multimedia today. By becoming more specific about what interactivity means, I hope to demystify the term a bit. In some cases, the qualities we value as interactive are not particularly new, although commercial publicity tends to portray the new medium as groundbreaking. Specificity helps us see the continuities between CD-ROMs and already existing media; vagueness encourages us to reiterate the hyperbolic advertising aimed at selling more "interactive" products.

It seems to be easier to describe what is *not* interactive. Interactivity, as structuralists would argue, is defined in opposition to other things, not in terms of any intrinsic qualities. Thus it is important to look at the kinds of media that CD-ROMs are clearly differentiated from.

CD-ROMs constantly position themselves as having qualities that you can't find in the dominant media of our time: film and television. Selling this difference, whatever it is, becomes a way of creating a new market and providing it with product. It is against the social context of the dominant current media that we should begin to read the concept of interactivity.

The discourse of interactivity can be read as a criticism of the perceived shortcomings of film and television; CD-ROMs, it is implied, provide what those media cannot. Inter*activity* is obviously meant to be opposed to *passive* ways of receiving media, and so this term is part of a long discussion about the "passive" watcher of film and television. The concept of passive viewing suggests that you don't have to do that much when you watch film/TV. You sit back in your recliner/theater seat and let the sea of images and sounds wash over you. The TV couch potato or the film viewer doesn't have to do much except stay awake in order to keep up with what's going on onscreen, according to this way of thinking. The movie or television program continues regardless of whether you are conscious and alert or dozing in your seat.

This understanding of the media is part of their appeal. If you want "mindless entertainment" at the end of a hard day's work, then television, conceptualized as a passive form, can provide the goods. Similarly, much of the appeal of many current Hollywood blockbusters is the pleasure of letting the movie take you on a "roller-coaster ride." You are strapped in and propelled through the narrative without much control over what will happen, which can provide thrills without requiring much activity on the viewer's part.

Yet this understanding of popular film and television as essentially passive pleasures has negative ramifications for those media. In a culture that values action, a passive art form is necessarily given lesser status. Watching hours and hours of television is something that many of us do, but a passive conception of television makes it seem like a guilty pleasure. We know that we really should be *doing* something, and since television watching is obviously "doing nothing," it must be a waste of time.

Interactive CD-ROMs are clearly *not* television and film, understood in this way. They do not proceed if the user dozes in his/her chair. They require physical and mental action from the viewer in order for the CD-ROM to proceed. Unlike film and television texts, most CD-ROM games come to a standstill if the player falls asleep. They require constant participation by the player, either at blitzkrieg speed (as in shoot-'em-up games such as *Quake*) or at a more deliberative pace (as in *Zork* and its progeny).

Although the physical action required for most games is minimal (moving a mouse or a joystick), it is significant compared to the couch potato's inertia. The couch potato's lack of physical activity is easily equated with his/her lack of mental activity, causing this figure to be vilified. The fantasy of interactivity in CD-ROMs positions the player as more mentally active than a couch potato, and the fact that the mouse/joystick user is more physically active echoes this assertion. Even the slightest degree of added movement that comes with using CD-ROMs can be read as a signal that we are progressing away from the couch potato syndrome. But if one steps aside from these rhetorical shadings, just how different is using a computer mouse from using a TV remote? The concept of interactivity emphasizes the distinction between mouse clicking and remote clicking, instead of noting the possible continuities between the two.

Another significant context for understanding CD-ROM interactivity is the discourse of "control" and the mass media. Widespread discussions of the effects of television violence on children, subliminal advertising, and media-related stalkings are predicated on the belief that television can influence behavior, often toward the antisocial. In these discussions television is a medium that is outside our control at the very least, and at most a medium that controls people, particularly the unstable or the gullible.

Interactive multimedia revels in the fact that you the user are in control. You can travel to any virtual space you want on the CD-ROM. You can stay there as long as you want, doing whatever you choose to do, according to the ideal conceptualization of CD-ROMs. You are behind the controls of *Microsoft Flight Simulator*, and so your commands propel the CD-ROM. In *Sim City* you have the godlike powers to create natural disasters and wipe cities off the map. CD-ROM games often offer the fantasy of extending our control to technology that far exceeds most of our real grasps: jumbo jets, interstellar cruisers, and laser rifles. They can offer a simulated feel for what it must be like to have control over a superbly trained body, such as a black belt karate fighter. While television can only offer control of what stations to watch, the CD-ROM promises the fantasy of control over an entire world.

Such fantasies call upon central ideological values in the hypercapitalist society, such as "choice." Having a "choice" is perceived as necessarily better, allowing us to exercise our freedoms. The concept of democracy enshrines choice as a basic principle, and capitalism has refined this rhetoric to make choice a fundamental principle of business. Markets should be relatively "free" because this will encourage more competition and therefore more choices for the consumer (and more choice is better). Choice is such a central value in modern America that the government rejected a national health plan largely because of an argument that it might limit Americans' choice of which doctor to see.

Commercial media also use the rhetoric of choice to sell their wares. Cable television, for instance, rose to prominence largely based on this rhetoric: more channels, more choices, therefore better television. The fantasy of five hundred possible channels offering five hundred different options seems to be the current goal of cable television's pursuit of choice. Yet CD-ROMs differentiate themselves from television by emphasizing choice. Television offers only the choice among different channels, but the activity remains the same: passive watching. Television may offer you "fifty-seven channels" but, as Bruce Springsteen sang, often the viewer finds there's "nothing on." Although cable systems emphasize the number of choices they present to viewers, CD-ROMs activate an implicit criticism of this kind of choice as being too limited, too monotonous.

The fantasy of interactivity offers a choice of different kinds of activities, not just one (watching). Interactive players can run, leap, punch, drive, fly, shoot, and fuck within the virtual worlds of the CD-ROM. Players have choices about where they travel within those worlds as well as choices about what they do when they get there. But there is a bit of slippage in this conception of the CD-ROM player's activity. It is our emissary to the virtual world who runs, leaps, punches, and so on; we the actual players are only clicking a mouse or moving a joystick.

The concept of interactivity, applied to current CD-ROMs, blurs these distinctions, emphasizing the range of choices presented to our computer avatars instead of admitting how few choices of activity they present to our bodies. Technological optimists can do away with such quibbles by saying that the mouse and the joystick are only temporary solutions that will be supplanted by true virtual reality gear. The hyperbolic rhetoric of interactivity, tied into the ideology of choice, makes it all too easy to substitute a possible conception of the future for a description of current CD-ROMs.

Although the description of film and television as essentially passive media still has great power in social discourse, in academic circles this conception has been overturned by decades of writing by media studies scholars. There are no such things as passive media, cultural studies scholars assert. The process of making meaning is an active one. Texts are not containers full of predetermined messages; instead they require us to complete the cues provided by the film/television programs, so that we make meanings for ourselves.

Ideology enables us to make some of these meanings. Based on a few cues given by a commercial (a smiling baby), we are able to summon a whole host of ideological labels (innocence, helplessness, etc.) to apply to this image. In this sense, making meaning is always active. Similarly, the structure of narrative (particularly classical narration) encourages our mental activity. By encouraging us to keep asking, "What's going to happen next?" classical narration creates expectations that we keep revising as we receive new information. When a beautiful young woman in a horror movie goes to take a late-night shower, we expect the worst to happen to her (based on our previous genre experiences), and a filmmaker can use our hypothesis to create tension. This image of the spectator hypothesizing, observing, and revising portrays the film and television viewing process as much more active than outward appearances might lead us to believe.[6]

Reader response theorists and scholars of fan behavior have emphasized the enormous range of meanings that actual viewers can assign to texts and the many uses they can make of a popular culture text. We are learning that fans can do many different things with a film or television text rather than just watch it from beginning to end, particularly if they take advantage of the capabilities of the videocassette player. People watch videos and turn them off before the ending. They rewind horror videos and watch the gory sections over and over again. They tape *The Simpsons* and watch portions in slow motion to look for jokes hidden among the images. They turn the sound down on *The Wizard of Oz* and play Pink Floyd's *Dark Side of the Moon* as a substitute sound track. They intercut Kirk's and Spock's lines from *Star Trek* to create a homosexual romantic scene. By learning more about what fans do with texts, we academics are learning just how "interactive" mainstream media can be. Not only do people "actively" make meaning from texts, but they also interact with texts in quite complicated ways to make new, reshaped conglomerates.

If almost all media require "active" processing and can be used "interactively," how can it possibly make sense to talk about "interactivity" as a defining quality of any one medium? If a person with two VCRs, some cable, and a stack of porno and *Star Trek* tapes can create erotic gay space fantasies, how is that so different from using Steven Spielberg's *Director's Chair*, a CD-ROM that lets you edit together clips from raw movie footage? How can we talk about the "interactivity" of a medium, considering the range of fan interactions?

It is instructive, I think, to differentiate between the things people do with texts and the practices a medium seems to call for. When you watch network television you are clearly encouraged to keep watching the same channel, lured away from the remote control button by never-ending promises of what's "coming up next." Similarly, a film in a theater clearly announces how you should watch it: by staying in your seat without smoking or talking. These metamessages about how to watch television or films communicate much to society about the nature of those media, and these messages powerfully shape how we conceptualize those media. Film announces that it should be watched with a certain physical/verbal passivity, which shapes the way we think about the movies.

So many of our actual media practices violate these metamessages that one is tempted to downplay their importance. After all, people do talk and move around in a movie theater, and people do channel surf through their cable systems. And yet in our social discourse about the nature of movies or television or CD-ROMs, these messages about what interactions the text seems to encourage are important. When I discuss with my students the practice of editing together a Spock and

Kirk romance, inevitably one of them will say something like, "But the creators never intended that to happen." And that student is quite right. When we read a text, we also gain a sense of how we're "intended" to read that text, and undoubtedly certain fan interactions with *Star Trek* texts violate those suggestions of what we're "supposed" to do with the episodes. At the metalevel it's clear that we're "supposed" to watch a horror video from beginning to end, although in practice actual viewers can fast-forward to the "good parts."

The distinction between what actual viewers do with texts and what the text seems to call for them to do is not a hard and fast distinction, since it is necessarily changing. If enough people engage in a media practice, that activity can become an institutionalized assumption that guides both viewers and producers. For instance, the practice of zipping through commercials has become so common that commercial producers have shifted strategies to encourage viewers to stop zipping and watch their particular commercial. The producers of *The Simpsons*, knowing that fans use slow motion to watch tapes, hide jokes in the program's minutiae for those viewers' pleasure. The nature of the activity that a medium calls for is constantly evolving, but that does not mean that we cannot talk about those activities in productive ways. We can only talk about film and CD-ROMs and interactivity and passivity at a particular point in time.[7]

At this point in time, CD-ROM interactivity is defined both by ideological fantasies of control, choice, and activity and by actual experiences of particular CD-ROMs. A central assertion in this volume is that we can best understand the medium by looking at individual instances, because it is through individual encounters with CD-ROMs that the fantasies and the realities of the medium collide (or interact, if you will). Slowly, piece by piece, close attention to individual CD-ROMs gives us a fuller sense of what interactivity is, and this book starts along this long process.

Interactivity with Objects

In the meantime, however, we still need an initial framework to discuss interactivity that is not too strongly tied to commercial hype or too steeped in unself-conscious ideology. What, then, is interactivity in CD-ROMs today?[8] It is clearly not the passivity, lack of control, and lack of choice that we associate with the "intended" way of watching film and television. Nor is it merely the fantasies of choice, control, and activity that the industry's public relations has emphasized.

Let us begin with one of the more influential early definitions of the concept: the version of interactivity that guided development at MIT's Media Lab (as articulated by Andrew Lippman). According to Lippman, interactivity is "mutual and simultaneous activity on the part of both participants, usually working toward some goal, but not necessarily" (in Brand 1987, 46). What Lippman envisioned as the model for interactivity was the conversation: two people, each one processing the words of the conversation at the same time, each able to interrupt the other, yet both proceeding in a mutually agreed on direction.

This means that it is possible to discuss the "thread" of the interaction, an organizing principle that shapes the direction of conversation, without necessarily specifying a predetermined goal. Even the participants themselves cannot predict the outcome of a conversation, although conversation is an organized give-and-take, not a random collection of non sequiturs. The "mutual, simultaneous" activities of both participants in a conversation make the conversation's actual course impossible to chart. Similarly, the actual "thread" of the interactions between a software program and an individual user should be impossible to predict.

To his credit, Lippman did not think of conversation as a stilted exercise in turn-taking, where two people alternate expressing themselves in complete sentences. Lippman is clear that a real conversation involves interruptions, and so interactivity involves more than just alternation between participants. While this definition has much to its advantage, there is a flaw at its core. When Lippman envisioned the conversation as a model, he thought of a particularly rarefied form of conversation. Conversation to Lippman necessarily assumes that both parties find the conversation interesting. By placing the metaphor of "conversation" at the center of his definition, Lippman relies on a fantasy of what conversation is: a rich and free interchange of engaged minds.

However, real-life conversations only infrequently live up to this standard. Real conversations all too frequently veer into directions that we do not like and cannot control. When my Aunt Sophie starts discussing the details of her lumbago, I can't run screaming from the room without causing a major family incident. Real conversations are often dominated by the mundane but unavoidable details of everyday living: who will pick up the children at school, or who should remember to bring home the tater tots from the grocery. In addition, real conversations occur within the structure of existing power and relational structures. Rarely do conversations occur between participants of equal power, and therefore one person often dominates a conversational topic, making the interaction rather one-sided. Conversations between me and my supervisors are limited by the power relations, which keep me from expressing my opinions of them freely. Lippman's valorization of conversation as a central model for interactive software ignores the all-too-real fact that we often seek out interactive games as a way of *avoiding* such conversation.

Part of the pleasure of playing a CD-ROM is that it is not at all like real conversation. If a CD-ROM interaction veers into territory I find too frustrating, I can shut down the program with no long-term social consequences. CD-ROM interaction promises pleasures that are usually far removed from the mundane. In general, I feel that I can control my movements within the virtual space of a CD-ROM—and when I can't control them, I usually find myself in interesting territory anyway. In real-life conversations, the opposite is all too often true. A real-life conversation is much less dependable than a good CD-ROM.[9]

Lippman's model echoes the nostalgic valorization of "conversation" that circulates in criticisms of popular culture. In the good old days before television, we are told, families engaged in rich conversation with each other about all kinds of topics. Before that, the same argument could be made about the good old days before radio. Somewhere in our mythic past, there was such a thing as good conversation. But in all likelihood, conversation had the same capacity for boring us then as it does now. The fantasy of conversation that guides Lippman may be powerful, but it is only tenuously rooted in reality.

Although it is perhaps easier to conceptualize interactivity as modeled on interactions with other people, this intuitive fit may do more harm than good. Instead I assert that interactivity is better conceptualized as interaction with *objects*, not with people.

This notion may seem counterintuitive at the start, since our interactions with objects seem quite one-way. Throwing a rock seems more like an "action" than an interaction. But the objects in a CD-ROM are no ordinary objects. They operate under a basic principle for multimedia design: any depicted object, whether an image of a person or a rock, is potentially capable of an enormous range of response to our actions. A rock thrown in the real world is propelled through the air toward a target. One might predict that if the rock is clicked on it will be hurled toward a target, but in multimedia you never know until you try it. A rock in a CD-ROM may respond to clicks with a giggle, an explosion, a fart, a trumpet fanfare, or a song-and-dance. CD-ROM designers are not bound by the technology to linking rocks to throwing; they can just as easily link rocks to rolls, or rocks to clocks.

Within the world of the designer, this capacity for response is made possible by what is called "object-oriented programming." Designers think of a person's interactions with the computer as a range of possible "events" that occur to "objects" the designer defines. The designer's job is to associate responses to events. If a user presses the escape key or clicks on a rock, those are events that call for responses by the program. Any response can be linked to any event. One could potentially link the help function to clicking on the rock or the rock-moving function to the escape key.

An important part of defining an object that is "mouse-able" is delimiting the object's logical boundaries. An object, logically defined to the program, need not be exactly the same as an object onscreen. One can make two logical objects (having two separate responses) out of what appears to be a single onscreen object (for instance, one can define a click on a sword handle and a click on the sword blade to provide two different responses). The fact that that area appears to be inhabited by a single onscreen object is not crucial. From a player's point of view, a CD-ROM can be inhabited by monsters and warriors; from a designer's point of view, it is composed of events and objects. When one defines a mouse-able object, one is really defining an area to be clicked on.

This way of thinking about programming fundamentally reshaped the structure of the CD-ROM's virtual space. In the real world, we expect that human agents are capable of a wide range of responses, but that objects have relatively limited responses to our actions. If I throw a rock I might hit a window or a police officer, but the rock won't make off-color suggestions or snicker at me. In the real world those capacities belong to humans alone.

What object-oriented programming does is to give objects and people/characters in a CD-ROM a kind of radical equality. There is no reason why CD-ROM people/characters should be able to snicker and objects should not. After all, both are nothing more than areas defined on the screen. It takes no more effort to make a CD-ROM rock snicker than it does to make a CD-ROM character snicker. In the world of the CD-ROM, "people" and "objects" are equally capable of responding.[10]

Lippman's discussion of interactivity as conversation makes sense if one assumes that only humans are capable of rich response. However, when you reconceptualize objects as being capable of initially unpredictable responses, this radically reshapes their possibilities for interactions.[11] People in conversation can be undependable in creating a rich conversation; objects can be programmed to be quite dependable in their interactions.

Once we stop comparing interactivity to a conversation between people, we shift away from Lippman's dependence on both parties in the conversation doing "mutual and simultaneous activity." The CD-ROM player cannot necessarily tell if the computer is performing a task or sitting in a wait state, nor is the player particularly concerned with the computer's processes. What a player is concerned with is: what am I capable of doing to the objects onscreen and what can they do to me (or my emissary to the virtual world)? What matters is the interface, not the processing that goes on behind the scenes.

Three factors shape our interactions with CD-ROM objects: what the player can do to objects onscreen, how the objects respond, and whether the objects can initiate actions.

What can the player do to objects onscreen? The obvious answer, in most cases, is that all the player can do is click on objects. But many games translate the click into different virtual actions. Moving the mouse to a particular area on the screen can change the mouse's function from a gun to a grabber that can hold things to a specialized function (such as striking a flame when the mouse touches a match onscreen). Often these different mouse actions are given different mouse icons: a target for the "gun" function, a hand for the "grabber" function.

In the real world we tend to have a wide variety of options available. I can use my hand to grab, to slap, to tickle, and to wave. Even tools that have a seemingly straightforward purpose (such as a hammer) are not limited to their primary use. For instance, in the real world I can use a hammer as a paperweight. But in the CD-ROM world, our range of possible actions toward objects is quite limited.

You can usually use a match only to light a virtual fire, not to clean your virtual teeth. The quality of interactivity depends on the possible actions we can perform on objects.

What then can the object do to respond to our actions? As mentioned earlier, a CD-ROM object, unlike a real-world object, is potentially capable of an extraordinary range of responses. Out of these seemingly infinite responses, which one(s) actually comprise an object's reaction? Are they predictable (does the monster explode each time I fire at it?), or do they vary? How many possible responses does an object have?

What can these objects do to us (or our avatar) without our prompting? Can they initiate an interaction? Can they prompt us to make a response? Can a monster attack us, or do we have to seek it out? A shoot-'em-up game like *Duke Nukem* depends on an enormous number of objects that initiate attacks on us. In a more leisurely game like *Riven*, objects rarely demand such immediate attention; instead, we have to search them out for interaction.

These three factors determine much about our interaction with the CD-ROM. Does the CD-ROM depend mostly on our finding the correct actions to drive the program (e.g., *7th Guest*)? Does the CD-ROM's pleasure come largely from the interesting/idiosyncratic responses we get to our actions (the *Living Books* series)? Is the CD-ROM driven by actions initiated by the program, demanding our response (*Duke Nukem*)? The text then can set up a cyclical interaction: our action, the text's reaction, the text's instigating action, our reaction to it, and so forth. Or the text can de-emphasize one portion of the cycle. For instance, many CD-ROMs depend on the action-reponse cycle (you click, it responds), rather than initiating an action without our bidding. Due to the nature of multimedia, CD-ROMs sometimes switch among these different interactive modes, and the CD-ROM critic must be sensitive to such shifts.

Interactivity with the Narration

The initial unpredictability of CD-ROM objects is the first fundamental that shapes the quality of our interactions with CD-ROMs. The other fundamental interactive possibility is the ability to interact with the narration. Narration is the way the story (the narrative) is told. A particular story can be told with a variety of possible narrations. Two different stage productions of a play can present different narrational strategies in interpreting the same text. Different viewpoints, order of presenting events, tones, and emphases in narration make for a different experience of the story.

The possibility of interacting with the narration creates infinitely more complexity. Not only can the same story be told in different ways, but interactivity presents the added possibility of changing the story itself. Interactivity with the narration implies more than just reinterpreting the same narrative; it also brings the possibility of new and different story events. Interacting with the narration allows CD-ROMs to reshape the events that make up the story.

Interaction with the narration can take several forms in CD-ROMs. First, CD-ROMs can provide several different perspectives on the action. For the film and television director presenting a scene, the choice of which camera angles to use and what order to present them is a matter of controlling information to the audience. Do we get to see the ticking bomb under the table, or does the film withhold this information? How early in the scene should the audience discover the hidden bomb? Do we see one actor's frightened face but not another's? The director's/editor's/cinematographer's jobs are made up of many minute choices of perspective to present, and each choice has a bearing on what information an audience has and when they have it. The process of giving and withholding information (largely done through controlling perspective) can give the film or television program suspense, pathos, or surprise.

Interactivity of perspective in a CD-ROM seems to imply that we have a broad range of choices in viewing a scene. Ideally, it would seem, we should be able to "walk" freely through a scene, choosing what perspectives we find most interesting at the moment, somewhat like a museumgoer browsing through a statuary garden. (I refer the reader to Alison Trope's chapter for more discussion of the relationship between museums and CD-ROMs.) We should be able to look at the various characters/objects from many sides, and we even should be able to look under the table to check for hidden bombs. In actuality, the limitations of CD-ROM memory capabilities mean that we cannot see all possible perspectives on the action, or even all desirable perspectives. The CD-ROM designer, like the film director and editor, must choose which perspectives to allow the player to have. What objects can we move all the way around? Can we look above, below, or inside it?

An added difficulty for the CD-ROM designer is that the player often expects to get a broad range of perspectives, particularly in more recent games. If we are given too few choices of perspective we feel that the viewpoints have been preselected for us (as in a film). When playing *7th Guest*, you have no choice but to be whisked through the scenery by steadicam camera movements. The designer places boundaries on what we can spend time examining. However, if the designer gives too many possible "camera angles" on the action, he/she potentially sacrifices the narrationally powerful option of withholding information from us. If the designer lets us see the bomb under the table, then he/she cannot dependably create a surprise explosion the way a film director can.

The range of viewpoints we are given, therefore, shapes our game play. If the CD-ROM "cuts to a close-up" of only those objects that will help us solve a puzzle, then we can be confident that most object close-ups will provide narratively significant information. This provides a different form of interactivity from a program that lets the player see a range of viewpoints, even those that contain no information crucial to puzzle solving. In such a CD-ROM, the attention to detail will be different, since one never knows which objects will be significant. The choice of which perspectives to offer the viewer may be based on the medium's memory restrictions, but it also affects the way we interact with the virtual space.

A second form of narrational interactivity involves interacting with the series of events. By "event" I mean a story occurrence that has lasting, irreversible effects on the game: solving a puzzle, picking up an object that will be needed later in the game, killing a character, and destroying or building a structure. Simply picking up an object (a wooden stake, for instance) is not necessarily an event; however, picking up and carrying a wooden stake you will need later is a story event. Your capabilities as a player have changed, as you will discover if or when you encounter a vampire. An event changes the game's status.

Narrational interactivity can allow us to choose which events will be important in our story and in which order we will witness those events. Early "interactive" print books offered the ability to choose which events made up a particular reading of a novel. Using a tree structure, the book presented a series of decisions for the reader to make, which influenced which events made up the story ("If you want Bill to call for help, go to page 45. If you want him to run away, go to page 61"). Interactivity with events at least opens up the possibility of choosing which events the story will comprise.

At present there are few commercial CD-ROMs that act on the early promise of tree-structure print narratives and hypertexts to allow different collections of events to make up the CD-ROM's "story." In most CD-ROM games, the player must witness/accomplish *all* narratively significant events. You must collect all the significant objects in order to escape the house, or you must successfully navigate all levels of play, or you must visit all important virtual sites. If you leave out any one of these narratively significant events, you cannot win the game (without the help of secret "cheat codes," which allow knowledgeable players in some games to bypass the restrictions). The interactivity of events presented in commercial CD-ROM games tends to remain fairly limited, at least in comparison with other earlier interactive texts.

The fear seems to be that if designers did not force players to complete all the tasks, the players might miss some of the spaces and puzzles the designers worked so hard to create, and so they would not get their "full money's worth" for the CD-ROM. However, if more CD-ROM games allowed us to select which events will make up our particular story (instead of requiring us to experience all narratively significant events), this might extend the playing life of the game. Rarely does a player revisit and replay a game such as *Myst,* which requires you to visit all the sites and solve all the puzzles before you "win." Why should they? All the information in the CD-ROM has been exhausted. But if the game allowed us to choose among particular series of events leading to different outcomes, we might be more inclined to interact with the game again to experience a new and different set of story events.

The possibility to choose different narrative events was publicized as an early potential of interactive media, but this possibility seems to have fallen by the wayside as developers create CD-ROMs. The reason for this, clearly, is economic. There is no fiscal advantage to encouraging users to revisit a CD-ROM like *7th Guest*. Commercial Web sites and television series encourage people to revisit them so they can continue to reach the consumer with their advertising messages. A

CD-ROM, however, is purchased once, and the producer receives no more money from the player who frequently revisits the disc than from one who does not. Therefore, the structure of interactivity currently dominant in commercial multimedia de-emphasizes an interactivity with events that might encourage revisiting the disc.

Economic factors, as mentioned earlier, are crucial in shaping our current conception of interactivity. However, we should not lose sight of the possibility for interactivity with events demonstrated in hypertexts and tree-structure books. New delivery systems may restructure the economics of the strongly designed object, and we need to remember the capacities that are neglected by the current economic system. If producers begin to put a metered charge on access time to a central game site, the economics would favor a kind of interactivity that encourages players to revisit the site over and over (such as an emphasis on interactivity with events). We need to recognize the way commercial forces shape our current understanding of interactivity so that we are not trapped into a single way of conceptualizing interactivity.

What today's commercial CD-ROMs do offer instead is the ability to interact with narration concerning the *sequence* of events. If we are required to visit all the sites on a CD-ROM, at least we are able to visit them in an order of our own choosing. This differentiates multimedia products from film and television texts, which clearly specify the order in which we see and hear things.

On first impression this ability to rearrange the sequence of events seems to be one of the primary differentiations between film/television and CD-ROMs, with CD-ROMs providing almost unlimited freedom to perform actions in a player-determined order. However, CD-ROMs, like film and television, also restrict and set boundaries on the sequence of actions. CD-ROMs can place restrictions on the sequence, requiring us to do certain activities before we are able to do others. Perhaps players can function in a cave only after they have successfully obtained a torch, or maybe they can find a secret passageway only after they have read a book left by the owner.

Such events in CD-ROM games tend to be contingent but not chained. A film/television narrative chains together contingent events to create a seemingly airtight sequence. If a man wants to solve his wife's murder, he must find the one-armed killer; to find the one-armed killer, he needs address information on amputees; to find this information, he must break into a hospital; and so on. Each of these events is contingent on previous events, and these contingencies are chained together so that any large failure jeopardizes the overall goal (to find a killer). In a CD-ROM there may be contingent events (you must find the torch before you can go in the cave), but failure to accomplish any one event does not necessarily bring the whole story to a halt. It merely prevents you from accomplishing the contingent event. There are usually plenty of things to do in the meantime: other sites to visit, other contingent events to accomplish.

The CD-ROM does not present infinite freedom to change the sequence of events. In fact, the restrictions that a CD-ROM places on your ability to change the event sequence are crucial to the quality of the interactivity. More restrictions tend to bind the narration toward a single story line, and fewer restrictions give a wider horizon of interactivity.

More interactivity is not necessarily a better idea. Restrictions create tension or give a sense of accomplishment. Part of the reason one feels pleasure when unlocking a particular puzzle in a CD-ROM is that you know it's not easy. You know you had to accomplish several other tasks before you solved the puzzle, and these contingent events are necessary to provide a sense of achievement. Commercial CD-ROMs tend to balance the emotional payoffs of restricted, contingent events with the implied ability to influence the overall sequence of events.

CD-ROMs, then, present us with the possibility of interacting with the narration by changing perspectives or rearranging the sequence of events. In addition, interactivity with the narration involves the possibility of selecting different goals. In most CD-ROM games, the overall goal is specified for the player. In *Phantasmagoria* the player must stay alive and send the serial killer to his bloody grave. In *Myst* one must solve the mystery of what happened among Atrus, Sirrus, and Achenar. The player has no choice concerning what goal to pursue. There may be other mysteries to solve on Myst Island or other supernatural phenomena to investigate in *Phantasmagoria,* but the player will never get a chance to investigate them. The goal has been irrevocably set by the narration; there is no opportunity for us to interact and change that goal.

In this area of interactivity, CD-ROM games lag far behind other kinds of CD-ROM programs. For instance, the genealogy databases that Pamela Wilson investigates in her chapter allow the "player" to investigate any number of questions. Such a database necessarily has to be open to a wide range of goals, since it is marketed to a broad audience seeking answers to many different genealogical mysteries. The goal for an interaction with a family history program is determined by the particular user. Although such CD-ROMs are not as flashy as the 3-D graphics and gore of *Diablo,* they are more interactive in the singular sense that they allow us a choice of goals.

CD-ROM games could learn from the interactivity in these other programs. Since CD-ROM games cannot duplicate the full complexity of a real world, they will never be able to provide an infinite story space with an unlimited number of story goals for us to investigate. However, allowing us to choose game goals may encourage players to revisit the virtual spaces. Knowing that other stories and other goals can be investigated on Myst Island might encourage me to return there. Just as amateur genealogists return to their CD-ROMs to pursue new goals, other CD-ROM players could be lured back to the games with the promise of pursuing alternate goals.

Once again, economic factors currently do not encourage this kind of interactivity, since there is no fiscal motivation to encourage viewers to revisit most CD-ROM games. The different economics of family history CD-ROMs encourages designers to emphasize a different kind of interactivity. Economics and utopian discourses interact to shape the ways designers create interactive discs.

Interactivity: Overarching Factors

In summary, the concept of interactivity may be broken down into two broad categories: interactivity with objects and interactivity with the narration. Interactivity with the narration allows us to change perspective, select and sequence events, and select goals. Interactivity with objects depends on what we can do to the objects, how they can respond, and what they can do to us.

These two sets of factors determine much of the nature of interactivity with a CD-ROM. Two other important, overarching factors affect the quality of both our interactivity with objects and our interactivity with the narration.

First is the number of options available to the player at any one time. As Lippman has noted, interactivity depends on the *impression* that we have an infinite series of choices available to us. This impression, for Lippman, is the difference between the selective and the interactive. In a selective game one has only a very limited sense of possibilities at any given time. For instance, early *Adventure*-type text-based games restricted the player to a few possible movements (up, down, right, left, forward, backward). Such games offered options, but in Lippman's terms they were selective, not truly interactive. By this definition the "interactive" books that allowed us to choose one narrative path over another are selective and not interactive. Interactivity gives the impression that we have many possible options at any given time. Of course a CD-ROM cannot present a truly infinite array of options, but it does not have to. To feel interactive, in Lippman's sense, a CD-ROM must only convince us that there *seem* to be infinite possibilities for the player's actions.

I do not wish to continue Lippman's labeling of certain programs as selective and others as interactive. The impression of infinite choices (as in many modern CD-ROM games) adds an interactive quality, but so does the ability to choose different plot events (as in interactive tree-structure books). To say that one is selective and one is interactive is too coarse a distinction. Instead we can say that they are interactive in different ways, that they emphasize different interactive factors. Discussing the specific forms of interaction helps us describe a text more precisely.

Most commercial CD-ROMs tend to give a sense of infinite choice by emphasizing our ability to interact with objects and (to a lesser extent) our ability to rearrange the sequence of events. As noted earlier, CD-ROMs often restrict our ability to change the sequence of events (in order to cause tension, provide a sense of accomplishment, etc.). They tend to counterbalance this restriction by allowing us to click on and interact with many objects, thus helping to give the game a sense of infinite possible options at any given time. The *total* number of possible combinations for clicking on objects or accomplishing events seems infinite, thus making more modern CD-ROM games seem more interactive than, say, a tree-structure novel. The multiple pathway novel has one kind of interactive advantage (the ability to choose different events), but it has less of the overarching impression that we have infinite options.

The other overarching factor that affects interactivity is the frequency of the interaction. Does the CD-ROM encourage fast-paced interaction, or does it tend to evoke a slower interchange of actions? Thus far our discussion on interactivity has emphasized the kinds of interactions between player and CD-ROM. The frequency of those interactions can be just as important a factor in the quality of our interactive experience. Both *Doom* and *Myst* are highly interactive texts, but one of the primary differences between them is the pace of the interaction. *Doom* actually offers a relatively limited choice of action: basically we can move in various directions, and we can shoot at demons using different weapons. There is less of the sense of infinite possibilities for action at any one time. However, because the demons come at us so rapidly, forcing us to respond, the interaction between player and object is highly charged. The pace of the interaction helps boost *Doom*'s interactivity. *Myst*, on the opposite extreme, tends to evoke a slower pace of interaction. Because few objects in *Myst* demand our immediate response (as the demons in *Doom* do), we interact with objects less frequently. To compensate for this lack of quick pace, *Myst* emphasizes the sense of having almost infinite choices of what to do on these islands.

Using these concepts, one can begin to describe the nature of interactivity with a CD-ROM more precisely. *Myst* offers no choice of goal and makes certain events strongly contingent on others (particularly concerning solving the puzzles), giving us a limited capacity to change the sequence of events. It offers a fairly large (though far from infinite) range of perspectives on objects, and the slow pace evoked accentuates our ability to manipulate certain objects. *Doom*, like *Myst*, offers no choice of goal. Unlike *Myst*, however, it offers a relatively limited set of actions toward objects. It counterbalances this limitation by the fast pace of objects (demons) acting toward us, provoking a series of rapid responses by the player. *Doom* also offers a fairly large set of perspectives on objects, but because of the breakneck pace, the breadth of perspective is relegated to the background of our awareness.

This is by no means a complete description of the games. Certainly there are many factors that contribute to our overall experience of the two games: the violent nature of the action and the horrific monster imagery in *Doom*, the detailed realization of environments and the complexity of puzzles in *Myst*. Yet these factors do not contribute to the *interactivity* of these games, though they certainly are important to the pleasure of games. Although interactivity is praised as being the primary payoff of playing a CD-ROM, it is by no means the only pleasure, and in some cases may not even be the primary pleasure. The emotional/cognitive payoffs of playing a CD-ROM also come from its interweaving of themes, tones, moods, voice (ironic, authoritative, etc.), and style. These factors play a large part in the pleasure we gain from a CD-ROM, although they are not necessarily interactive qualities. The overwrought horror style of imagery in *Phantasmagoria* is crucial to the experience of the game, even though such stylistic concerns can be viewed independently of its interactive qualities. To explain a CD-ROM's interactivity is not to explain the entire CD-ROM, although publicity emphasizes the medium's interactivity.

Interactivity is only one quality (although an important one) that shapes our experience of multimedia. We as critics need to be able to separate the hype about interactivity from the structure of CD-ROM texts, recognizing what pleasures and frustrations come from interactivity and what experiences come from other factors. Learning how to describe the nature of these games' interactivity is a useful starting place for the real work of analysis: examining the details of the text or of the particular interaction between a player and a CD-ROM text.

The Essays

The first essay opens up many of the issues that will be examined throughout this anthology. Janet Murray and Henry Jenkins examine several *Star Trek* CD-ROMs, discussing the tension between the interactive potential of the technology and the commercial desire to target a particular audience. Using a mixture of close textual analysis and ethnographic inquiry into a well-known fan community, Murray and Jenkins look at the differences among corporate-designed *Star Trek* products and other electronic fan practices (such as *Trek* chat rooms), and they argue over the forces that constrain and facilitate certain kinds of fan interactions.

My essay looks at another attempt to adapt a text with a devoted fan community into a successful CD-ROM. I use *Monty Python and the Quest for the Holy Grail,* the CD-ROM adaptation of the midnight movie classic *Monty Python and the Holy Grail,* as a case study to examine certain questions. How does one translate the linear medium of film into the nonlinear medium of multimedia? Are there certain kinds of films that are more easily adaptable? If so, what qualities might make it easier to translate a film to CD-ROM?

After dealing with science fiction and parody adaptations, this anthology turns its attention to horror. Angela Ndalianis investigates how *Phantasmagoria* borrows from modern horror films. Beginning with current film theories on audience engagement with slasher films, Ndalianis discusses how these theories of interpretation should be adapted to discuss the more fluid interactions called for by nonlinear horror texts.

Brian Kelly and Scott Bukatman discuss how particular CD-ROMs reconfigure imagery of a different sort. Kelly and Bukatman discuss the fascination with predigital machinery that characterizes much of digital culture. In an essay whose topics range from windup toys to cutting-edge CD-ROMs, they resituate the CD-ROM as machine, as an interface involving both bodies and memory.

With Ted Friedman's essay, the anthology begins to examine how CD-ROMs transform our understanding of space and place. Friedman looks at how *Civilization II* encourages us to "think like a computer," giving us a new perspective on narrative, time, and space that enables a map to become the hero of the story. But underlying this radical restructuring of textual interaction is a far-from-radical ideology of nationalism and imperialism, as Friedman details.

Alison Trope explores how various CD-ROM technologies have encouraged a reconceptualization of the idea of the museum. Since a CD-ROM museum is not limited by the physical constraints of walls, it can play with the notion of what a museum is. Discussing a range of traditional and nontraditional CD-ROM museums, Trope examines how these CD-ROMs negotiate the tension between the utopian possibilities of the technology and the constraints of working within the contexts of cultural institutions. Surveying museological discourses, she reminds us that CD-ROMs are always produced within an institutional context, not in a universe of pure aesthetical potential.

Pamela Wilson discusses how CD-ROM and Internet technologies are combining to change the concept of the archive. Examining the practice of family history, Wilson discovers a postmodern industry that combines major software producers (such as Brøderbund) with amateur genealogists. Her ethnographic research shows how these technologies are being used to establish a new kind of "kinship," making them important in the identity politics of a seemingly rootless era.

Vanessa Gack also uses an ethnographic approach to her research. She studies two children who are participating in an after-school program entitled the 5th Dimension, which includes a computer gaming environment. Closely examining the way these two children interact with games, with each other, and with other children, Gack discusses different strategies of mastery these children demonstrate. Situating the gaming environment within the already existing social environment, she looks at the ways computer games continue and alter social relations.

Finally, we come to the space that is most familiar to academics: the classroom. Leslie Jarmon, perhaps the first person to have a dissertation accepted in CD-ROM format, conveys her experiences in using CD-ROMs in academic environments. A strong advocate of the advantages of CD-ROM technology, Jarmon discusses how new technologies have revolutionized the way she teaches. Lisa Cartwright also shares her experiences in teaching a class that asked graduate students to submit their final scholarly project on hypermedia. For Cartwright, this experience was a decidedly mixed bag. Although several students produced interesting projects, Cartwright reminds us of the institutional factors that make teaching such classes difficult.

Altogether these essays demonstrate the advantages of sticking closely to CD-ROM texts, the contexts in which they are made, and the contexts in which they are used. This close attention gives a portrait of both the shortcomings and the potentials of CD-ROM technology today.

NOTES

1. For discussions of *Afternoon, a story,* see Moulthrop (1989) and Bolter (1991).

2. In addition, this delay causes media scholars to lose a great deal of primary information about the reception of texts when they must deal with archival material instead of available contemporary sources. This makes the history of the medium difficult to write because there are few trained scholars examining the phenomenon as it occurs.

3. Or perhaps CD-ROMs will follow the path taken by VHS videotape. Having gained a broad chunk of the market, VHS remains the dominant

videotape format in spite of numerous other challengers that provide better picture and sound. Perhaps CD-ROMs, like VHS, can provide enough satisfying functionality to keep consumers from feeling they need to purchase more hardware.

4. Similarly, Noël Carroll (1985) argues that much of the power of movies is attributable to the fact that they organize their imagery with "economy, legibility, and coherence . . . to a degree not found in everyday experience" (93). The movies have developed devices to focus our attention on narratively crucial phenomena, encouraging us to ignore tangential information, which gives the movies an urgency not found in most daily life.

5. Some CD-ROMs (such as *Doom*) allow the player to compete against other players online, blurring the straightforward distinction between online interactivity and strongly designed objects. However, I believe that it is more useful to consider online *Doom* play to be more like the experience of interactivity in a strongly designed object. What a player is experiencing is the thrill of hyperviolence designed into the *Doom* universe. The fact that you may be playing someone in an office across the country certainly adds to the overall experience of playing *Doom*, but it does not change the basic *interactivity* of the experience. You may have more (personally) at stake knowing that you're competing against an actual human being, but the game itself is not radically different. As far as the interactivity of *Doom* (and other online/CD-ROM games) is concerned, it matters little whether you're playing a person or the computer. What changes in these instances has less to do with the strongly designed object and more to do with the framework for interpreting the game play.

6. This discourse, like that of interactivity, can best be understood in its context. The concept of "active" reading was created in opposition to the dominant understanding of media consumption as passive. The use of the word "active" in this sense is an attempt to give value to the viewer's mental processes, instead of continuing to devalue them as "passive." The term "active" contains more rhetorical force than it does analytic insight. Few would disagree that we add ideological shadings in interpreting imagery or that the expectations and anticipations are central to our screen experiences. To realize that viewers do such activities is not particularly surprising, at least not as surprising as hearing these processes called "active" in the context of a "passive" conception of media. Like the uses of the word "interactive," the academic use of the word "active" is both truthful and an exaggeration. We do engage in such complicated mental processes sitting in front of the screen, but few who consider these media to be "passive" would deny that these processes are engaged. By labeling these activities "active," academics say less about the viewing process than they do about the value placed on that process.

7. If one discussed media at the height of the nickelodeon, one might have discussed film as a highly interactive medium, since the nickelodeon was a place where audiences verbally and physically interacted with each other and with the screen. Until the exhibitors tried to attract a more genteel audience than its largely immigrant clientele, there were no rules about sitting down and being quiet in the movie theaters, and the interchanges created a significantly livelier experience than the one we now associate with the movies.

8. In discussing CD-ROM interactivity, I will lean heavily on examples from CD-ROM games because games tend to make the richest use of interactive capabilities. However, games do not make the best use of all interactive factors, and in these cases I refer to non-game texts. This introduction is about strongly designed CD-ROM texts, not about CD-ROM games alone, although the market is so driven by games that these products dominate our evolving understanding of what interactivity is.

9. Of course online chatting (in chat rooms, multiple user dungeons/domains [MUDs], e-mail discussion lists, and online bulletin boards) has been credited with reviving the long lost art of conversation, but these forums are nowhere near interactive in the way Lippman describes. They are turn-taking affairs (chat rooms often rely on signals such as "otu"—"over to you"—to show that one has finished a thought). One can meet interesting people on the Net that one never would have encountered elsewhere, and the anonymity possible in chat rooms allows people to try on races, genders, sexual preferences, and so forth in ways that are fascinating to explore. These online conversations, however interesting they are interpersonally and theoretically, do not rethink the structure of interaction. They primarily allow you to extend your conversational realm outside a conventional distance-bound sphere and to explore more flexible, alternate, multiple subjectivities within those conversations.

10. This argument about the equality of persons and objects in multimedia echoes a similar argument about the nature of silent film (and particularly silent film comedy). Neither humans nor objects were allowed to "speak" in a silent film; this took away a primary advantage that human characters generally enjoy over inanimate objects in fiction. Because characters in a play can speak, they are more likely to engage our sympathies than onstage props are. The realm of silent film put people and objects on equal footing, which might help explain the omnipresent struggle between humans and recalcitrant objects in silent film slapstick.

11. When one first clicks on an object, its response is unpredictable, but thereafter the object tends to respond in a robotic, repetitive fashion. Most objects in a CD-ROM, when clicked on, give only one response, which you can repeat over and over again by repeated clicking. There are exceptions where single objects respond in multiple ways to multiple clicks, but the

memory constraints of the CD-ROM do not always encourage such multifunction programming. As these constraints loosen, I expect multimedia to explore further the possibilities of objects responding in multiple ways to our interactions.

WORKS CITED

Arnheim, Rudolf. 1957. *Film as Art*. Berkeley: University of California Press.

Bolter, J. David. 1991. *Writing Space: The Computer, Hypertext, and the History of Writing*. Hillsdale, NJ: Lawrence Erlbaum.

Brand, Stewart. 1987. *The Media Lab: Inventing the Future at MIT*. New York: Viking.

Carroll, Noël. 1985. The Power of Movies. *Daedalus* 114, no. 4 (fall): 79–103.

Moulthrop, Stuart. 1989. Hypertext and "the Hyperreal." In *Hypertext '89*, 259–67. New York: Association of Computing Machinery.

Stone, Allucquere Rosanne. 1995. *The War of Desire and Technology at the Close of the Mechanical Age*. Cambridge: MIT Press.

Turkle, Sherry. 1995. *Life on the Screen: Identity in the Age of the Internet*. New York: Simon and Schuster.

Chapter Two

Before the Holodeck

Translating Star Trek *into Digital Media*

Janet Murray and Henry Jenkins

Over the past three decades, *Star Trek* has offered viewers a succession of compelling and ever more sophisticated models for the future of digital media, starting with the voice-activated computer and working through the holodeck, the holosuite, and most recently, the holonovel. The various *Star Trek* series consistently depict the ways a future culture will work and play within digital space, as Picard dons a trenchcoat, Data pits his detective skills against Moriarty, Bashir saves the world as a James Bond–like superspy, and *Voyager's* holographic medical program does battle with Grendel. The holodeck and its descendants represent an immersive and fully interactive environment, which allows ship crewmembers the chance to enter into fantasy environments, assume fictive roles, and escape from the mundane reality of always having to go where no one has gone before.

And, of course, as regular viewers can tell you, every time anyone uses the holodeck, it crashes or malfunctions.

This last feature may be its strongest commonality to the digital media currently on the marketplace. Contemporary CD-ROM games, as consumers regularly complain, take forever to install, are full of bugs, and offer only limited interactivity. Compared with the technological wonders of the holodeck, we are, to borrow a memorable phrase from the original series, working with "stone knives and bearskins."

We might, however, view the situation a bit more optimistically. Imagine that we are publishing this collection in 1895 and that only last year the Lumières produced their heart-stopping image of a train arriving at La Ciotat Station. Would we have been able to imagine *Birth of a Nation* or *Children of Paradise* or *Pulp Fiction*? Suppose we are writing this essay just a few years after the advent of broadcasting. Could we imagine works with the narrative complexity of *Murder One* or the cultural influence of *Star Trek*? The emergence of electronic media tempts us in a similar way: can we, looking at the gamelike and stick-figure narratives of the 1990s CD-ROMs, glimpse the emerging genres of a more expressive narrative art? Will our current moment be treated by future historians as the prehistory of the holodeck, just as film historians once published books called *Before Griffith* or *Before Mickey* (Crafton, 1993; Fell, 1983)?

Aesthetic Potentials

In *Hamlet on the Holodeck: The Future of Narrative in Cyberspace*, Janet Murray (1997) identifies the defining properties of this emerging medium as encyclopedic, spatial, kaleidoscopic, participatory, and procedural. That is, computers contain vast amounts of information; they can be navigated as virtual spaces; they are made up of parts that can be juxtaposed, rotated, and rearranged to form varying gestalts; they call forth our actions and respond to them; and they are built out of rules of behavior that allow us to animate the worlds we model. The current computer environment is often referred to with the additive term "multi-media," but to the extent that designers exploit the intrinsic properties of the computer, they forge the component parts (including text, still images, moving images, graphic design, the drawn, and the photographed) into new media-synthetic representational forms capable of capturing the world of our senses and the world of our dreams with an astonishing new immediacy.

The aesthetic pleasures that emerge most immediately from the intrinsic properties of the computer medium are often referred to as immersion and interactivity. Immersion is the pleasure of being transported to another place, of losing our sense of normal reality and extending ourselves into a seemingly limitless, enclosing, other realm, where we move and act under different and often magical rules. Interactivity includes simple participation, common to other aesthetic experiences such as call and response singing, and also the pleasure of agency, the power to act freely and to make choices and to have those actions and choices affect the narrative environment. Immersion and interactive agency reinforce each other. The more we feel we are surrounded by another enticing environment, the more we want to manipulate it. The more the environment responds to our manipulations, the stronger our involvement with it, and the more persuasive the illusion of being there.

If we are to imagine the computer growing into a locus for a maturely expressive art, we would expect to see storytellers building upon these qualities: making imaginary worlds with ever more encyclopedic completeness, more navigable spaces, more capable of exploration from multiple points of view, more open to our active participation, and displaying increasingly complex behaviors that are both surprising and appropriate. If we believe that electronic media can grow in these ways, we might begin to imagine a future art of immersive and interactive narrative worlds that would have an expressiveness comparable to that of contemporary film or television.

How can we judge whether such hopes are justified? How can we gauge our distance from an artform that does not yet exist? One way is to measure existing electronic narratives against a comparable experience in the nonelectronic world. At the present moment, a high percentage of the digital media on the market are second-order phenomena, adaptations of texts that gained their popularity through film and television. In a horizontally integrated media industry, characters, plots, images move fluidly across various media, participating in what Marsha Kinder (1991) has called the entertainment supersystem. We need to develop an equally horizontally integrated media studies that can trace the migrations of these stories and understand what is gained and lost as they move across different media. The *Star Trek* "franchise," as Viacom increasingly calls it, represents perhaps that premier example of a transmedia phenomenon, having developed five television series, eight feature films, a succession of best-selling original novels, comic books, toys, and various other spin-off products.

Fans as Interactive Consumers

Star Trek represents an important test case for digital media's "interactivity" because we already have established a complex picture of the program's varied audiences, its active fans, and their current interactions with the program materials (Harrison et al., 1996; Bacon-Smith et al., 1992; Penley, 1997; Bernardi, 1998). From its initiation, *Star Trek* sought to provide a richly detailed narrative universe that could be the platform for many different characters and situations and thus attract diverse audiences. *Star Trek* might one week offer us a mystery or a scientific problem to be solved, another week a combat situation or a diplomatic crisis, or still another week a political allegory. For some fans the scientific and technological challenges of the twenty-fourth century form the central focus. For others it is the close social bonds between the vividly portrayed characters. For still others it might be the program's utopian vision of a future founded on intergalactic cooperation and fellowship, on the IDIC and the Prime Directive.

In *Textual Poachers* (1992) and *Science Fiction Audiences* (1995), Henry Jenkins offers detailed accounts of many different interpretive communities, each with its own interests, preferences, and expectations about the program. For example, he asked a group of male MIT students and a group of female fanzine writers to tell what came to mind when he named each character. For the technologically inclined males, what seemed most salient were the capabilities of the characters as autonomous problem solvers, stressing instances where one or another, primarily acting alone, saved the ship from certain destruction. For the female fanzine writers, characters could not be discussed alone but were understood within a web of character relationships; they stressed the centrality of the characters' emotional lives, desires, and motivations to the experience of the series. The stark contrast between these two ways of thinking about the characters reflect broader, gender-specific approaches, which in turn shape the different fan communities and their evaluations of the episodes.

The program producers have become increasingly adept at balancing those various expectations, shifting the weight given A-plots focused on military and technological issues and B-plots focused on character and cultural relationships from week to week or from series to series. Such negotiations allow the cagey producers to hold the interests of a fan community that, according to a Harris poll, counts its membership at more than 50 percent of the American public. In turn, the multiple fan communities surrounding *Star Trek* have long appropriated the characters, props, story lines, and ethos of the series as a participatory narrative. Fans elaborate on the program "metatext" to insure both consistency and complexity in their understanding of its various worlds and cultures, imagining alternative plots or different character relations.

Fans have been on the cutting edge of new technologies, using the photocopier, desktop publishing software, and the VCR as creative tools for interacting with and reworking the aired episodes. Electronic environments offer a vital new venue for this participatory and immersive fan culture. Cyberspace already hosts numerous news groups, Web sites, and e-mail lists facilitating exchanges between program fans. At a recent fan convention Henry Jenkins attended, a great deal of discussion centered around how fans were using online communications to expand their sense of participation in a larger fan community as well as the potential disruption this has caused to face-to-face interactions since many fans still lack Net access.

As Murray argues, the properties of electronic media clearly address the active fan's desire to become more directly engaged with the enticing narrative world. For instance, because computer environments are encyclopedic, they can encompass all the information the fan could wish in one enormous network: guides to all the episodes, megabytes of stills, dictionaries of alien languages, textbooks on twenty-fourth-century physics, and so forth. Because computer environments are intrinsically spatial, they can embody imaginary locations in navigable form. The *Enterprise* can be seen on the television or movie screen, but viewers can control their movements through the ship only on their computer screen. The computer can contain a more densely realized world than live-action costumed role playing because it can incorporate the *Star Trek* mise-en-scène. Users can assume the role of the actual *Trek* characters, as portrayed by the series actors, and direct their action. In short, it promises to make the imaginary world more realized and more participatory than ever.

On the other hand, as Jenkins argues, this interactive technology contains the potential of controlling and regulating fan pleasures at a time when Viacom has increasingly sought to centralize and constrain grassroots fan activities. The encyclopedic nature of digital *Star Trek* reference works reopens the question of what information is valuable and what is marginal. Commercially available programs construct a rigid barrier between canonical information contained in the aired series and apocryphal information originating in the fan community. Interactive fictions facilitate certain fan pleasures (especially those most fully compatible with the male-centered genres of current CD-ROMs and video games) at the risk of excluding altogether other pleasures (especially the focus of the female fans on character relations and alien cultures). The medium exerts its own pressures on the program materials, often resulting in a decreased focus on character (which is still difficult to realize in the digital realm) and a greater emphasis on action (especially tactical combat, puzzle solving, and navigation). While television's *Star Trek* must appeal to multiple demographic groups to sustain its popularity, the digital market is still predominantly young, male, and technically oriented; the content of the *Star Trek* games reflects that orientation, using technology to facilitate interactions with fictional technologies, and emphasizing combat over conversation.

Moreover, simulation games are frequently ideologically coercive, naturalizing their core assumptions about the way the universe works into procedural knowledge needed to perform well in the game; we are "programmed" to think within certain parameters, while the participatory structure of the game allows us the illusion of free choice. The ideology of the TV programs is constantly disputed among fans, but *Star Trek* games necessarily take sides. The result is a strikingly militaristic conception of the *Enterprise,* its mission, and its relations to alien cultures. The games also display paternalism, as aliens are most often cast as subjects requiring rescue and assistance, and misogyny, as powerful women are often cast as dangerously duplicitous. No one seems to have hardwired the Prime Directive into these simulated *Trek* worlds.

None of these limitations are intrinsic to the technology; rather, they reflect the current state of the game marketplace and the limitations of corporate understandings of the potential *Star Trek* audience. We both remain optimistic about the broad potential of this new medium to express many different fantasies and facilitate diverse interactions. However, there is a danger with any new medium that the codification of generic and ideological conventions at its founding moment will have a lasting impact on how the medium gets used, who it speaks for, and what kinds of stories it tells. In our excitement about the potentials of the new encyclopedic medium, we need to be attentive to what it excludes as well as what it includes; in our fascination with technological interactivity, we need to be aware of the social and cultural interactions it refuses to facilitate. As you read this essay, you will find traces of the tensions between the two authors. Murray expressed greater enthusiasm for the ways these games exploit the potentials of their medium, while Jenkins was more dissatisfied with their failure to facilitate the desires of diverse fan communities. We hope that acknowledging these disagreements will make us more conscious of the competing interests and expectations shaping the emerging medium.

Sampling Digital Star Trek

Focusing on the applications surrounding the *Star Trek* world allows us to make a taxonomy of the narrative experiences currently available in electronic environments and to measure them by the standard of complex, immersive, participatory engagement already enjoyed by *Trek* fans through other means. Specifically, we will be looking at Simon and Schuster Interactive's *Star Trek: The Next Generation: Interactive Technical Manual;* Sega's *Star Trek: The Next Generation: Echoes from the Past;* and Spectrum Holobyte's *Star Trek: The Next Generation: A Final Unity*. These electronic environments allowed us to enter Captain Picard's ready room, put the *Enterprise* into warp drive, fire photons at Romulan warbirds, and beam down to exotic planets to face dangers, rescue the innocent, protect errant scientists and ecologically endangered alien wildlife, and struggle with whether to intervene in local political disputes. The three games represent three genres of electronic storytelling—the navigable virtual space, the skill-and-kill video game, and the exploratory puzzle game. We assessed how each exploits the properties of the medium and rewards diverse fan interests.

All three use *Star Trek*'s familiar theme music and visual design to establish their authenticity. They all take us to the bridge of the *Enterprise* as the central location and carefully detail its well-known stations. The *Technical Manual* uses the voice of Jonathan Frakes as Commander Riker for a narrated tour of the ship and the voice of Majel Barret as the ship's computer for more impersonal information about the various spaces and technologies. *Echoes from the Past* offers us graphics of all the crewmembers, and *A Final Unity* offers better character graphics and the voices of seven series regulars (who recorded more than a hour of dialogue each). In all three, the operation of the transporter with its tingling, whooshing sound offers the delightful sense of experiencing a magical environment firsthand. When Riker's offscreen voice in the *Technical Manual* pooh-poohs the fears of those who hate transporter travel by calmly explaining the (ludicrous) science behind it and exclaiming, "What could be safer?" he solicits our participation in the program world and rewards our familiarity with series history.

All these games successfully re-create the familiar *Enterprise* environment, suggesting the central role of the computer in creating illusory worlds. The more television and films rely on computer-generated effects and sets, the more we can be offered electronic versions of those same realities. For any narrative fantasy environment, the world exists partly on the stage, page, or screen and partly in our heads. It is, in the psychologist D. W. Winnicott's phrase, a transitional space, both external and internal, and therefore fraught with promise and desire. Furthermore, as Sherry Turkle makes clear in *The Second Self* (1995), we experience the computer itself, even without narrative content, as a similarly "evocative object"—both an external thing and an extension of our consciousness. To load this evocative space with powerful story material is to make it all the more compelling.

All these games build on our existing belief and emotional investment in the *Star Trek* universe. They also build on the complex metatext of program information viewers have constructed through their previous interactions. We come to the *Technical Manual* with knowledge of events that have happened aboard the ship. We come to the games knowing that if there is a warbird on the screen it may soon be attacking us and we had better put up our shields and get ready to fire our photons. The games spend little time explaining to us what phasers or tricorders are and how they work. We are primed for immersion and interaction when we hear the theme music.

The Technical Manual

The *Technical Manual* is one of the most detailed and navigable virtual spaces available in the game world, and combined with its evocative content, one of the most immersive. The manual was the first commercial product to use a technical innovation in PC-based virtual reality, QuickTime VR, developed by Apple Computer. QTVR can establish spaces we move through smoothly, as if viewing a camera pan, rather than step-by-step as if viewing a sequence of stills. Unlike *Myst* players, who can only step through separate, overlapping, but discontinuous separate drawings of the imaginary world, visitors to the *Enterprise* do not find themselves disoriented by missing pieces in the navigated terrain. Movement is circular in pan-like continuity or by jumps that provide either a zoomlike close view or a new positioning at another pan-able location. Some longer jumps include clips of the intervening space, insuring player orientation. The effect is a navigable, photorealistic, continuous space. The ambient sound shifts with each change of room. The detailed environment and easy mobility offer a strong sense of the *Enterprise* as a realized ship that persists when no one is watching the series.

The naming of the CD-ROM a "technical manual" and the inclusion of the Riker narrated tour point to a confusion of metaphors. A technical manual tells about an object in an abstracted sense, focusing on its design specifications and performance parameters; a tour is a visit to the object, which brings with it the expectation that we are going to see it in use in a specific context. Computers are appropriate both for collecting information—making the mother of all technical manuals—and for making visits to virtual objects. But this is a visit to a data bank rather than to an operating ship. What we expect but do not receive is a "fully functional" *Enterprise* we can take on adventures. Our tour of the *Enterprise* is more like a visit to Monticello or the U.S.S. *Constitution,* where we can look but not touch. We can see where famous events once occurred, but nothing is happening at the moment. Our tour guide keeps us moving along, not dawdling behind or taking unauthorized actions.

We can retrieve information and go "behind the scenes" at a privileged and glamorous location. We can experience the ship as a space to be moved through and examined, with extraordinary attention to the consoles and monitors we pass. There is even some effort to include some personally meaningful objects associated with each character. There's the pipe that Picard learned to play or his bound volume of Shakespeare, "items that give him much needed perspective on his duties"; there are Data's Mondrian-inspired artworks and his cat's collar and dish. But we cannot manipulate any of the technical wonders, cannot fly the ship or send Riker through the transporter. The designers have translated the technical data into a collection of text and images, rather than into a set of procedural rules that would support our interactions. The game casts us in the role of "authorized Starfleet visitors or trainees," instead of the heroic figures the series allows us to imagine ourselves to be.

Riker's narration consistently points toward the important role of human agency in the ship's operation. When the tour begins, he explains, "After being aboard the *Enterprise* for any length of time, you quickly come to think of it as a living entity. . . . Don't be fooled into thinking that the computer really runs the *Enterprise*. During command alert every critical operation is in the hands of trained Federation personnel." Such passages simultaneously provoke our technophilic fascination with the advanced technologies being represented and resolve our fears about our own ability to maintain control in the digital realm. However, such reassurances are ironic since the ship we see is so devoid of human agency. QuickTime VR is cleverly presented in the user documentation as "a subset of holodeck technology." But because QT is video-based and QTVR accomplishes its impressive illusion of continuous movement through a fixed set of video images, the space is static and depopulated.

The comparison with the holodeck foregrounds the narrative impoverishment of the *Technical Manual*. There is no present-tense story in this world, only pointers to stories from the series in a kind of archeological layering. In fact, there would seem to be three distinct levels of narration, all evoking narrative content only in the past tense: Riker's tour of the ship offers a personal picture of what it's like to live and work on the starship; the ship's computer offers a more impersonal narration, which describes the sanctioned functions and capabilities of the space, without any sense of the characters or events that occurred there; an accessible reference text may fill in aspects of public history—the commissioning of the various versions of the *Enterprise* as they have unfolded across the series, brief summaries of Klingon mythology, and so forth. For example, the computer stresses the fact that 10-Forward was designed for "optional socialization conditions," while Riker more evocatively speaks of it as a place where crewmembers "relax and unwind from the day's cares. . . . [or] hash things out over a drink." The printed text explains a sculpture found in Worf's quarters in terms of "the legendary struggle between Kahless the Unforgettable and His Brother, Morath, after Morath lied and brought shame to his family," while Riker speaks of Worf's internal conflict between his personal ties to the Federation and the Klingon empire.

The shifting levels of narration involve a complex play between public and private experience, history and gossip, routine activity and high adventure. None of these stories are presented in a compelling or immediate fashion; rather, these objects function as a high-gloss memory palace evoking our familiarity with the program history. The most personal of these levels, which includes references to such things as Data's struggles to understand the human realm, includes more focus on the issues of character psychology than one would anticipate from a technical manual. Yet if we see signs and artifacts associated with these characters, we never get to meet any of them. Riker's warm, sociable narration is the only detectable life sign.

The capaciousness of computer environments raises what Murray calls the "encyclopedic expectation," the sense that everything one wants will be available. For instance, we want all the personal objects in the crew's quarters to lead to revelations that deepen our sense of the characters' presence. But emotionally suggestive props, like the painting on Crusher's wall, remain unannotated and unidentified, contrasting sharply with the elaborate diagrams and technical explanations of the transporter or the holodeck. Similarly, we want to access the complete history of Starfleet, but the ship's computer somehow has only limited coverage ("We assume that other Federation starships have had histories as illustrious as that of the *Enterprise,* even though Kirk's ship seems to have become the most famous") and is similarly unable to document other alien races and cultures.

Much of fan cultural creation stresses the "realism" and completeness of the program universe, filling in missing information, resolving contradictory data, contextualizing episode events in the characters' lives and their cultures' history. The *Technical Manual* has not succeeded in incorporating even the information contained in the aired episodes, let alone offering a world as complex as the shared universe fans have created. The "encyclopedic expectation" aroused by digital media makes us more pressingly aware of the large gaps between what the viewers want to see and what the producers decide to give them. Who decides what to include in the encyclopedia? The range of narrative involvement provoked by the *Star Trek* universe suggests that a truly encyclopedic presentation of the official franchise world could sustain a broad range of interpretive communities, each looking for an inhabitable sector in the program universe. But the electronic *Technical Manual*, like its print predecessor, focuses on the information desired by young, technically oriented males rather than facilitating other reading formations.

Echoes from the Past

The *Technical Manual* offers a sense of realized space at the expense of interactive agency. It is a participatory environment, it is a persuasively detailed environment, and it is profoundly disempowering. The Sega Genesis game *Echoes from the Past* poses the opposite problems. Its graphics are crude and unconvincing but it has numerous opportunities for interactive agency. It is structured as a fighting game with exploratory aspects. The player can manipulate the familiar crewmembers, picking Picard's dialogue with alien trespassers on Federation space, navigating the ship from one planet to another at the CONN station, engaging in battles at the tactical station, distributing resources to life support, shields, and weapons at the engineering station, choosing away teams and beaming them down in the transporter room. There are also computer sensors to indicate if planetary atmosphere is safe and a database of technical information from the *Star Trek* universe, including background information that makes for a more realized game (the culture of the Romulans) and data on the immediate game world (which planets are habitable). There are many things to do, and most have clear and immediate consequences.

Playing the game means responding to a relentless series of red alert sirens. "This is way different from the show," complained one of our fifteen-year-old researchers. "I'm sick of everybody fighting us all the time." The *Star Trek* ethos certainly includes elements of adventure, conflict, even combat and overt intervention, but it is not reducible to the conventions of a fighting game. For many fans, the primary goals of the *Enterprise* continue to be exploration of unfamiliar space, diplomacy with unaligned cultures, and the furtherance of scientific knowledge, not, as the *Technical Manual* characterizes it, "the implementation of Federation Policy throughout the Galaxy." The fighting game offers few chances for meaningful negotiation across cultural, political, or racial differences. Any player who slowed down to think through the niceties of the Prime Directive will get blasted out of orbit. The *Star Trek* ethos doesn't shape our actions here; its iconography simply provides set decoration and pretext, sometimes not even that, as when our activity on one away mission consists of gathering up various puzzle pieces and reassembling them as the far from compelling means of getting a downed computer back on line.

When we are preparing an away team, the transporter room offers a broad array of anonymous red shirts and crewmembers who might join the series regulars: we choose between racial, gender, and physical types. Of course, even this element of choice is deceptive. In practice, character matters little in this game. Most Sega characters are simply iconographically enhanced cursors, defined not by their personalities but by their functionality, their capacity for action, though *Echoes* does little to differentiate between the characters in this regard. We do not enjoy being in an environment in which it makes no difference whether Data or Dr. Crusher holds the phaser.

Fan culture constantly reads series actions in terms of a deeper and deeper understanding of what they mean to the characters. The Sega video game, on the other hand, restricts the characters' behavior to a narrow repertoire of actions. Characters cannot learn from their actions in an environment where all choices are relatively arbitrary, unmotivated by what we already know about the characters, their motives, and their abilities. Characters become simply interchangeable parts. Even the technology is sometimes stripped of meaning. The characters gather up phasers and tricorders like so many gold coins or abstract tokens that contribute to the player's score but not toward resolving their plot predicament. The game genre, then, imposes too much on the *Star Trek* program materials, rendering our knowledge as regular viewers irrelevant. *Star Trek* becomes a brand name rubber-stamped on the package, an iconographic gloss on the same old game functions, not integral to the pleasures of participating in this virtual environment.

In its most successful example of agency and immersion, the Sega game allows us to move the ship to various warp speeds at the simulated CONN station. The series' presentation of warp speed, the rapid movement of stars outward from a central vanishing point, is a computer-generated effect. We are in the appropriate domain to experience space travel; setting the speed and destination gives the effect a satisfying immediacy. The CONN simulation also employs temporal duration to add a level of realism often lacking in the series itself: flying from point to point at a slower warp speed takes longer and means that the player must wait for the next phase of action, but flying at a higher warp speed carries risks and may not be feasible under certain conditions. Simulations are often most interesting in the limitations they place on player action. On the other hand, this resistance becomes frustrating if it seems too arbitrary and mechanistic, if the restrictions on our actions do not seem plausible in terms of the fictional universe.

The *Technical Manual* and the Sega game illustrate opposite failings. The *Manual* unimaginatively sticks with linear formats (the illustrated manual, the guided tour) instead of inventing a new form for the new medium (e.g., a test drive of a functional starship). *Echoes*, on the other hand, takes a narrowly conceived computer-based form and tries to shoehorn the resistantly rich *Star Trek* material into this format.

A Final Unity

A Final Unity is a puzzle game, a better fit for the *Star Trek* world than the hack-and-slash video game, because much of the series involves figuring out puzzles through scientific (or pseudoscientific) reasoning. The fit between the task and the ethos of the series increases our narrative immersion. *A Final Unity* is closer than the other two texts to the narrative engagement of the television series for other reasons as well. Both the scenario design and the dialogue are comparable to the television episodes in quality. The graphics are also closer to the video images, though the characters still have a puppettoon stiffness, simplified facial features, and elongated foreheads. (The realism of the story space and the abstraction of the human figures perhaps reflect the narrative priorities governing the game design process.) The characters' dialogue is spoken by the cast members, including all the principal players.

The pacing feels more like a series episode, with entrances, exits, and much discussion. This is both an advantage and a disadvantage. It makes the story more compelling and vivid, but it slows down the pace and often reduces the interaction to stepping through a scene, clicking the mouse as each character speaks. There is a long expository period without much opportunity for significant choice. Character interaction is minimal, and completely related to plot development. The game runs us through a succession of problem-solving adventures with few consequences for the characters. Dialogue is reduced to the uncovering of salient information. The effect is like watching a whole season of A-plots, back-to-back. Worse yet, plot demands force characters to behave contrary to their psychology, as when the independent Worf repeatedly advises to check with Starfleet before taking actions.

A Final Unity uses its superior memory and computing power to present better graphics, lots of audio, and a more extensive set of player actions. Patrolling the Neutral Zone on alert for Romulan transgressions, the *Enterprise* rescues a scout from attack by a Garidian warbird. The player can delegate the fighting to Worf (who will win, of course) and thereby bypass the battle game in favor of more plot. This design decision fosters immersion and interactive agency. The player feels accompanied in the adventure, as if playing alongside the actual crew, and can also adapt the game to the individual tastes in activities. One player may work through complex engineering problems another relegates to the Geordi surrogate or may choose to maintain control over astronavigation or medical equipment. The game can be as technical as its MIT student fans prefer or may shortcut much, though not all, of the technological questions.

After the battle's successful completion, the captain interviews the three people beamed aboard. Depending on whom Picard asks, the crew was engaged in an archeological expedition to reclaim a lost scroll of deep spiritual significance for their people or was looking for an ancient text that will spark political reform on their planet. During this phase, the player has limited control over the activity, choosing only the order of people interviewed. The player does not even determine what questions to ask or lines to deliver, since it is necessary to ask all the questions in order to gather the information needed to effectively play the game. Player control comes only in sequencing dialogue, not in shaping the tone and nature of the interaction. This participatory design offers no true sense of agency.

Strikingly, this elaborate exposition is immediately dismissed. We are repeatedly hailed and lured to alternative (and seemingly unrelated) rescue missions. We move through space in the helter-skelter manner of a picaresque adventurer. The first hail is from survivors of what seems to have been a Romulan attack on a scientific station. An away team is chosen for us and beamed to the surface. A maze and puzzle game replaces the bridge interface, which by now has become irritatingly monotonous.

The puzzle game involves an icon-based command structure, which keeps us actively engaged. In the main window, the away team is at the doomed scientific station. Below it are a series of iconized controls: a set of faces to toggle through in order to choose the active character; a set of icons that can replace the cursor to initiate commands: a foot for moving, a hand for using, an eye for viewing, a balloon for speaking. We also have a tricorder, medical tricorder, communicator, medkit, and phaser at our disposal; we can pick up other useful objects (with the hand) as we walk through the multiroom world. The commands are explained by a text display linked to an active cursor; as you pass a foot over an area, if it is a place you can walk, the words "walk to conduit" will appear. This is a smooth way of allowing diverse activities, while letting the player know which ones are legal and which are not. More irritating are the repeated spoken messages when you make an illegal move, as when Riker paternalistically lectures us, "I don't think that would be a wise course of action" or Geordi whines, "It doesn't work."

Having seen Geordi perform engineering miracles on the series, we find it particularly satisfying to operate his character, sending him to a table full of gizmos or a control panel or to talk to an engineer. The other characters are given large enough audio parts and distinctive enough characterizations to make the dramatic world persuasive. Significantly, most problems we encounter require the expertise of Geordi, who functions as a surrogate for more technological viewers, while Crusher's skills as a doctor are largely disposable. At one point in the game, an alien spacecraft repeatedly phasers various crewmembers, and they are knocked unconscious by the blasts. No one races to their rescue, they receive no medical attention, but they pop back up with no sustained injuries after only a few moments. On the other hand, a failed power conduit requires a complex process of locating and applying a succession of high-tech tools.

One key pleasure of this world is using the familiar *Star Trek* paraphernalia. For instance, we can save a woman pinned under a large pipe by giving its coordinates to the transporter. It is fun to see the object slowly appear in the transporter, accompanied by the familiar transporter chimes. The tricorder is also usable and can be consulted to discover important information that contributes constructively to the mission. But after a while the pleasure of using this paraphernalia pales. Dedicated players can tolerate tremendous repetition in fighting games and can climb and fight their way through maze after maze as long as the spaces display small variations between progressive levels. But in a puzzle game we want the worlds to look truly different from one another and for the people we meet to have the particularity and interests of dramatic characters. Yet, despite the trappings of more interesting plots, involving diplomatic disputes, missing naturalists, matriarchal cultures, planets deciding to enter the Federation, and so forth, this puzzle game still reduces the plot to a series of interactions with the mocked-up *Star Trek* gadgets.

In fact, many of the more character-based elements actively retard the game structure and frustrate more action-oriented players. The Web page we used as a crib sheet for the game dismissed the character-driven conversations as "talk, talk, talk." Such hostility to character interaction runs through Net group responses to the aired episodes, often directed against female characters. This perspective gets reinforced by the severe limitations in the game's ability to construct and communicate character relations. Often we can see the same exact lines emerging from two or more characters in response to the same situation, as if it really made no difference who was communicating the information or issuing our instructions. Often the same lines are repeated throughout the game in a variety of situations as the characters seem unable to respond to their changing environment. If the game designers decide that the character has nothing to contribute to a given problem, they provide them with no dialogue. Many times, Picard urgently asked his second-in-command for advice and was told, "I have no thoughts on that matter at this time," a response that amounts to either insubordination or malperformance of duties. If they are impatient with such "talk, talk, talk," game players can display dialogue in text and click through it more rapidly.

The range of choices of actions (walking, looking, speaking, and using) and objects in the various game worlds creates a frame of mind more suitable to free play than to the constrained domain of the game. Brenda Laurel (1993) would argue that computer-based narrative environments should always be more like children's make-believe than like constrained games. This is particularly hard to do with franchised characters. In the *Final Unity* game, when we tried to use the command system in a constructivist way, we came up against two sets of restrictions: the limitations on story conventions in the *Star Trek* universe and the limitations on moves in this particular world. When our young research assistant tried to phaser Riker, the response was only a bland "That would not be the wisest thing to do." A more interesting response would have been to kill or injure the character and force the player to make do without his skills and input. When you try to disobey a Starfleet order or refuse a Starfleet request, the game stalls. Worf rolls his head and glares, Riker taps his foot impatiently, until we do as we are told. Yet the creation of a fictional universe as a play space, rather than as a board game, invites these subversive and aggressive impulses. Allowed into the story world, we want to rearrange it, test its limits. We want the increasingly deranged Riker to take the mole carcass we found on one planet and leave it in the captain's chair as a practical joke, anticipating how Jean-Luc might respond to his mischief. That the plot remains inflexible, no matter what subversive actions we may contemplate, makes it more like the professional novels (which are restricted from changing anything that might impact on the story options available to the television and movie producers) than like the participatory fan culture.

One wonders whether we will soon see a group of fan programmers who rescript the CD-ROM games in order to make them more perfectly satisfy fan interests, just as we have seen fan writers expand the stories and fan video makers appropriate and re-edit the footage. Already the same ingenuity that produced the composite videotapes that created love affairs between Kirk and Spock is being applied to editing digitized images of *Trek* characters. Will it be long before alternate episodes start appearing in cyberspace, or before fans themselves start compiling operable versions of the *Enterprise* and hijacking it for their own adventures?

MUDs as Participatory Media

When Jenkins spoke to women at a recent convention of fan writers and editors, most of them showed no interest in these computer games, even though many of them were active in online chat groups or in MUDs. These women were not hostile to the digital realm, but they preferred greater creative license and more networked communication with other fans rather than the games' preprogrammed structure. In the *Star Trek* MUD, TrekMuse, founded in 1990, over four thousand players have created their own characters. Five hundred people were enrolled in the Starfleet Academy in this MUD in 1995. Using typed commands and real-time typed communication among players, TrekMuse members can build their own virtual environments, talk to one another both in character (IN: Please reconsider your ultimatum) and out of character (OOC: Just because you're a Klingon, you don't have to be a jerk!). They can engage in explorations, battles, and diplomacy, apprentice to more experienced officers, and generally collaborate on a detailed, collectively imagined *Star Trek* universe.

Much of the MUD's activity involves treaty negotiation and diplomatic maneuvering. This is a common pleasure of live-action role playing, and in *Star Trek* worlds it is deepened by the concept of the Federation as a utopian organization. Of course, players are free to be greedy Ferengi or warlike Cardassians, but the world's ethos promotes justice and compassion for victims of oppression. One player, for instance, likens her identifications with the Bjorns (victims of brutal Cardassian oppression) to her sympathy, as a concerned Jew, for the fate of the Palestinians.

TrekMuse permits an appropriation and reworking of the general conventions of the *Star Trek* world. MUDs are an enticing narrative playground rather than an amusement park ride. Closest to fan culture in origins and philosophy, MUDs are perhaps the most promising environment for providing immersive and interactive experiences that combine the representational power of the computer with the free creativity of the fan culture.

The limitations of MUD narratives are related to their strengths. The stories that are made there are not as shapely or as satisfying narratively as the aired episodes. Because the world is improvised by a diverse and diffuse group, it is hard to create plots that require centralized planning. There are a limited number of workable formulas for collective action: the negotiation of treaties, the making of battles, the observation of protocol, the awarding of promotion, the arrangement of marriages. The pleasure of this kind of fiction is tied to the lack of closure in the environment. It is an open-ended world in which roles can be played until they lose their interest. Many players spend more time fleshing out their characters, imagining their backstory and their motivations, than in planning plots or solving technical problems. For many female fans of the series, the appeal is precisely in conversations, exploring similarities and differences with other players, and enacting roles that may be radically different from those they play in everyday life. MUDs and other online role-playing games create a sense of a populated world, distanced from one's physical self and obeying rules that are known and shared among the players.

Like larger fan culture, online *Trek* worlds extend the possibilities of the narrative without being bound by canonical events. MUDs facilitate the diverse interests of alternative reading formations, providing a space for both more character-intensive play and more combat-centered action. They bring to life things that are only referenced by the series. That five hundred people are enrolled in Starfleet Academy, which has courses, a faculty, and a system of graduation, is an impressive feat of collective imagination. It exemplifies a degree of immersive belief and an exercise of interactive agency that no top-down game could duplicate because it was conceived and executed by the players. Similarly, the creation of characters in a role-playing game has a satisfaction that goes beyond the mechanical control of a character taken from the series; the characters players choose reflect something of their own personal interests and fantasies. Each player brings to the game something of their own *Star Trek*.

Although the two authors agree about the pleasures and limitations of the games we examined, we came away from our limited survey with somewhat different perspectives.

For Jenkins, it is crucially important to note that most commercially released *Star Trek* digital artifacts target technologically inclined males. The focusing of media product on the interests of specific demographic groups is a logical outgrowth of contemporary capitalism even if it runs counter to the more polysemic strategy *Star Trek* adopted on television. As such consumers continue to dominate the market for games and other CD-ROM products, Viacom licenses only digital products that reward characteristically male reading competencies and interpretive interests, ignoring a more diverse fan demographic. A succession of such marketing choices will have a long-term impact on the development of digital media. The interplay of different audience demands pushes and stretches the media to accommodate a broader range of narratives and, in turn, enhance and enlarge its formal possibilities. A narrower audience not only threatens the commercial interests of the media producer (closing off potential consumers) but also stifles the medium's maturation. In the case of the *Star Trek* games, market strategies, ideological norms, genre conventions, and technological limitations interact to marginalize characters and their relationships and with them, the female fans who have regarded characters, rather than technology, plot, or mise-en-scène, as their primary entry into *Star Trek.*

On the other hand, Murray (who, though female, enjoys the technology, the combat, the characters, and, perhaps most of all, the richly imagined and hopeful futurism of the series) argues that we should resist the tendency to judge the potential of computer-based narrative by its current instantiations. We need to imagine alternative uses of the technology, alternative genres that might evolve, alternative solutions to its apparent limitations. Some subgenres, like fighting games, may remain frozen at a low level of narrative development, with satisfaction coming from other sources. But if we are seeing the birth of a powerful new representational medium, we can expect that the more successful elements of these diverse genres will coalesce and grow stronger. The immersive space of the *Technical Manual*, the feeling of interactive agency from taking the *Echoes* craft to warp speed, the integration of complex story structure and satisfying puzzle-based game interactions in *A Final Unity* are all encouraging indications of the possibilities of story-driven gaming. Most of all, the energy that active audiences are bringing to the world of MUDs holds the promise of greater participation not just in the machinery of the story world, but in the shaping of character and event.

As we go to press with this essay, a new area of conflict is emerging in the ongoing tug-of-war between *Trek* fans and *Trek* producers. More than fifty Internet sites run by fans were closed in 1997 when Viacom threatened legal copyright infringement sanctions. Viacom offers its own official site, claiming cyberspace as an above-ground medium, like television, rather than an underground medium, like photocopied zines. Fans have responded by organizing a resistance campaign. Meanwhile, television and the Internet are rapidly converging, as digital TV and rapid networking make their way into American homes.

The tensions we witnessed in these early games will be magnified as television struggles to open itself to the participatory story forms of the digital environment.

As we confront the rapidly emerging new digital narrative, we need to balance our optimism about its formal possibilities with a cautious concern for the commercial and ideological imperatives that are shaping it. As critics, we need to be alert to what is possible, and also to what possibilities are neglected. Though we may imagine a mature electronic narrative world with visual immediacy, high interactive agency, strong story lines, and open-ended constructivist architectures, a market-driven, male-controlled, and too often cynical industry may not be able to deliver it. The storytellers of the coming century may or may not be able to generate a medium with the generic diversity and broad range of pleasures embodied by the holodeck, which can fulfill Janeway's or Riker's needs for romance, Picard's hunger for mystery, and Worf's desires for combat; they may not even be able to match the satisfactions fans currently enjoy simply by trading linear stories with one another. Only time will tell. However, we can say with some confidence that the inventive energy and participatory momentum behind the *Star Trek* world will continue to push electronic narrative forms as far as they can go.

WORKS CITED

Bacon-Smith, Camille, Kent A. Ono, and Elyce Rae Helford. 1992. *Enterprising Women: Television Fandom and the Creation of Popular Myth.* Philadelphia: University of Pennsylvania Press.

Bernardi, Daniel. 1998. *Star Trek and History: Race-ing toward a White Future.* New Brunswick: Rutgers University Press.

Crafton, Donald. 1993. *Before Mickey: The Animated Film, 1898–1928.* Chicago: University of Chicago Press.

Fell, John. 1983. *Film before Griffith.* Berkeley: University of California Press.

Harrison, Taylor, Sarah Projonsky, and Kent A. Ond, eds. 1996. *Enterprise Zones: Critical Positions on Star Trek.* New York: Westview.

Jenkins, Henry. 1992. *Textual Poachers: Television Fans and Participatory Culture.* New York: Routledge.

Jenkins, Henry, and John Tulloch. 1995. *Science Fiction Audiences: Watching Doctor Who and Star Trek.* London: Routledge.

Kinder, Marsha. 1991. *Playing with Power in Movies, Television and Video Games:*

From Muppet Babies to Teenage Mutant Ninja Turtles. Berkeley: University of California Press.

Laurel, Brenda. 1993. *Computers as Theater*. Reading, MA: Addison-Wesley.

Laurel, Brenda, Rachel Strickland, and Rob Tow. 1994. Placeholder: Landscape and Narrative in Virtual Environments. *Computer Graphics: A Publication of ACM SIGGRAPH* 28, no 2: 118–26.

Murray, Janet H. 1997. *Hamlet on the Holodeck: The Future of Narrative in Cyberspace*. New York: Free Press.

Penley, Constance. 1997. *Nasa/Trek: Popular Science and Sex in America*. New York: Verso.

Turkle, Sherry. 1995. *The Second Self: Computers and the Human Spirit*. New York: Simon and Schuster.

Games

Star Trek: The Next Generation: A Final Unity. Spectrum Holobyte, 1995.

Star Trek: The Next Generation: Interactive Technical Manual. New York: Simon and Schuster Interactive, 1994.

Star Trek: The Next Generation: Echoes from the Past. Sega, 1993.

Chapter Three

"To Waste More Time, Please Click Here Again"

Monty Python and the Quest for Film/CD-ROM Adaptation

Greg M. Smith

In spite of all the ballyhoo about CD-ROMs as "new media," many commercial discs contain a high percentage of material recycled from other media. These CD-ROMs can mix together preexisting characters, clips lifted from videos and films, and game technology to create an interactive product that feels both new and familiar.[1] This recycled material provides a CD-ROM with a crucial economic advantage in the multimedia market: built-in name recognition. Consumers may not have heard of the latest original CD-ROM game, but they probably know *The Lion King* or *Toy Story* in their film incarnations. For corporate giants like Disney, these CD-ROMs provide another way to squeeze out more merchandising dollars from current successful properties. For those media properties that have exhausted their life on broadcast/cable/video, CD-ROMs provide another opportunity to extend their profitability. To feed the desire for content for this new medium, multinational corporations often can merely reach for their own shelves.

Spin-off CD-ROMs do not necessarily make the fullest use of the new medium's capabilities. They often merely place icons from the old medium (e.g., Simba from *The Lion King*) into familiar game settings or provide multimedia trappings around the presentation of movie clips (which the viewer must watch without interaction). It is all too easy to neglect the importance of these overt products of corporate synergy and to assert that the future of the medium must lie in innovative texts such as *Myst*. However, these commercially successful products shape the way we conceptualize what a CD-ROM does. They experiment with balancing familiarity and novelty in a way that can provide instant commercial appeal and long-lasting interactive rewards to their player/owner. For commercial multimedia, this combination of old and new is the much-sought-after Holy Grail.

The trick is "creating a CD-ROM that actually enhances its source material—and then shoots off in rewarding new directions." This is how *Entertainment Weekly* described 7th Level's *Monty Python and the Quest for the Holy Grail*, the CD-ROM it named the best multimedia product of 1996. *Monty Python and the Quest for the Holy Grail* (hereafter referred to as *Quest*) has been heralded as an exemplar of how to recycle material from other media while taking fuller advantage of the CD-ROM's interactive capacities. This essay examines the strategies *Quest* uses to transform its source material. What characteristics of the original 1975 film *Monty Python and the Holy Grail* make it particularly adaptable to multimedia? How does a successful CD-ROM spin-off expand on its source? In addition, I will demonstrate how *Quest* satirizes the notions of logic and productivity that have become underlying assumptions for both computer games and the real world of work in a cybersociety. *Quest* revels in its assertion that CD-ROMs are a waste of time in a world all too geared toward efficiency and time management.

There are other issues at stake here besides a further appreciation of the aesthetics and social commentary of Monty Python and 7th Level. As media corporations continue to mine narrative film and television texts as sources for CD-ROMs, the difficulties of translating a linear narrative into a multilinear structure become more apparent. All Hollywood films are not equal in their adaptability to the new medium. The first half of this essay begins to enumerate the qualities that might make one film more appropriate than another as a source for CD-ROM adaptation. Close examination of *Quest*'s successes can shed light on possible future interactions among film, television, and multimedia.

Sketchy Quotes

The troupe Monty Python began as a collaborative effort among experienced comedy players. John Cleese, Graham Chapman, and Eric Idle began their comedy work in revues at Cambridge, and Terry Jones and Michael Palin were writer-performers in the Cambridge cabarets. The various group members (including the American Terry Gilliam) sojourned in British television sketch comedy programs such as *Do Not Adjust Your Set* and *At Last, the 1948 Show* when they met to form Monty Python in 1969. Together the Python members created *Monty Python's Flying Circus*, an innovative comedy smash that ran on the BBC. The troupe's four seasons on television defined a particular comic style that came to be called "Pythonesque."

Pythonesque humor jumbled together bits and pieces of sketches in rapid succession. Densely verbal studio sketches would be followed by "something completely different": highly visual filmed slapstick, Gilliam's surreal animation, or stock footage of an audience of older women applauding. The Pythonesque involves a strange crossbreeding of low and high culture, an irreverent combination of farts and philosophy. They would simultaneously parody television shows and satirize intellectual pretensions (two of their favorite targets) in sketches such as the one that asked game show contestants to summarize Proust's novels in fifteen seconds. Python sketches would place stock characters in absurd situations that gave rise to escalating mayhem; for instance, a guidance counselor who encounters an accountant who wants to be a lion tamer, or a robber who mistakes a lingerie shop for a bank. Python members wrote these sketches and performed almost all the roles (including some memorable drag performances), which helped give them international celebrity after the program was syndicated. The Pythonesque blend of high and low, satire and parody, the absurd and the intellectual became a distinctively recognizable comic style around the world.

The international success of *Monty Python's Flying Circus* made it possible for the troupe to enter the world of filmmaking. Their rapidly assembled first film, *And Now for Something Completely Different* (1971), cobbled together bits from the first two seasons of *Flying Circus*. It was not until 1975 that the Python troupe created a film composed of original material done in the now familiar Pythonesque style.

The film *Monty Python and the Holy Grail* (hereafter called *Grail*) recycles the age-old tale of King Arthur, the Knights of the Round Table, and their quest for the Holy Grail (the chalice Christ drank from at the Last Supper). After receiving a charge from God, Arthur searches the land for knights to help him find the Grail, and soon they separate to pursue the Grail individually. Present are such familiar figures as Sir Galahad and Sir Lancelot, although their stories have been made indisputably Pythonesque. Galahad the Pure is tempted by a castle of maidens who appear all too eager to be punished by spanking. Lancelot, in his attempt to rescue someone he believes to be a damsel being forced into marriage, gets "carried away" and slaughters much of the wedding party before he discovers that the "damsel" is an effeminate man named Herbert. The Python film adds new players to the legend, such as Sir Robin, whose lack of bravery is proclaimed in minstrel song ("when danger reared its ugly head, he bravely turned his tail and fled").

The 1996 CD-ROM adaptation of the film is itself structured as a quest.[2] The player ventures among virtual sites patterned after locations in the film, such as the Plague Village and the Castle Anthrax. Visitors at the sites can participate in the familiar scenes by clicking on characters and objects. If you successfully navigate a site, you are rewarded with an excerpt from the film that takes place at that location. The player needs to collect various objects from the sites that will enable him/her to cross the Bridge of Death to receive the final reward: not the Holy Grail, but a scene excised from the original film.[3]

What qualities of the original 1975 film are particularly well suited for adaptation to multimedia? Part of the answer lies in the film's unusual structure. *Grail* follows the exploits of one knight and then another, heralding the new protagonist with a portentous voice-of-God narrator (explicitly announcing "The Tale of Sir ———") and trumpet fanfare. In so doing, the film follows the picaresque structure of the wanderings of Arthur and his knights. Instead of one long linear story, the legend of the Grail quest is composed of a series of individual stories. These stories need not necessarily be told in any particular order (other than the charge to Arthur, which initiates the quest, and Galahad's final discovery of the Grail). Although *Grail* can hardly be considered a "faithful" retelling of the myth, it does rely on the time-honored structure of separate, individual stories placed in the overall framework of the quest.

In this sense, *Grail* is an exception to the Hollywood norm. The classical Hollywood film tends to present a single chain of interlocking events bound together by a highly efficient type of linear causation. The film places the protagonist into a predicament and then presents his/her actions as he/she strives toward a goal. The spectator is constantly urged to ask, "What will happen next?" both at the local level ("Will the bandit get away in this car chase scene?") and at the global level ("Will the protagonist ever find the one-armed man who killed his wife?"). Classical film narration also aims to present its audience with little or no extraneous plot information. We can rest assured that if a mainstream film shows us a scene, the information in that scene will be necessary for us to make sense of the protagonist's pursuit of a goal.

To call such a structure merely "linear" does not accurately capture the taut structure of classical film. Hollywood films present linear stories, yes, but so do continuity comic strips (such as *Dick Tracy*) and soap operas. Hollywood films are constructed not only to present events in a particular linear order, but also to convince us that this is the only particular order in which the events could be portrayed. Classical narration gives a sense of inevitability to the particular arrangement of plot events. By promising to present a series of cause-and-effect actions with little extraneous material, the classical film makes it difficult to envision the story world as having alternative possibilities (other character qualities, other choices of action).

The classical Hollywood structure is so effective that many students find it difficult to conduct thought experiments in rearranging a film's plot structure ("Imagine how *Forrest Gump* would be different if it didn't start on the park bench"). This sense of inevitable plot order (and the expectation that everything we see will present crucial relevant information) makes the Hollywood film form particularly difficult to reshape. It is not merely difficult because Hollywood film is "linear"; it is difficult because most Hollywood films encourage us to believe that one could not leave out or rearrange anything without disturbing the whole.

Because *Grail* owes more to earlier picaresque storytelling forms than it does to tautly linked Hollywood structure, the separate stories in the film become more easily adaptable to nonlinear media.[4] Since it makes little difference whether we hear Sir Robin's tale before we hear Sir Galahad's story, this helps free the designers of the CD-ROM to consider these events independent. The designer need not worry about severing the narrative thread between two stories since the original film has already bracketed these tales off from each other. Galahad's tale, for instance, can then be located in a single virtual site on the CD-ROM (the Castle Anthrax). The CD-ROM does arrange these sites in a suggested order, which the player can follow by clicking the arrow pointing right, but players can also visit and revisit the sites in an order of their own choosing. Only one site has a specified place in a sequence (the Bridge of Death is the final obstacle, since it can be crossed only after one has collected and redistributed several objects). Constructing a CD-ROM as a collection of sites that can be visited in roughly any order is now characteristic of the medium. *Grail*'s return to a picaresque structure of separate tales makes the transition to discontinuous CD-ROM sites easier.

But only some of the film is structured as overtly framed tales. In fact, most of the scenes appear without the fanfare, in a way more typical of Hollywood films. These plot occurrences give the appearance of classical scenes, but they can be more accurately thought of as sketches. Understanding the distinction between sketch and scene is important for understanding what makes the original *Grail* film so adaptable to nonlinear media.

Much of the audience for *Grail* was already familiar with Monty Python because they had seen *Monty Python's Flying Circus*. The *Flying Circus* was structured as a series of sketches, following in a long tradition that extended from the British music hall to the stage revues of Oxford and Cambridge to Spike Milligan's television shows *Q5* through *Q9*, all of which influenced Python's particular version of sketch comedy. A sketch is a kind of scene and therefore tends to take place in a continuous block of time and space, thus betraying the theatrical origins of both sketch comedy and the cinematic scene. As we discussed earlier, Hollywood film *scenes* are interdependent units arranged into the highly interlocked chain of classical narration. They are rarely shown without some framework establishing the scene in the diegesis as a whole (necessitating talk show guests to provide a verbal "setup" before showing a clip). A *sketch*, however, is freestanding. It is an independent unit that can be placed in any of several different places in a vaudeville-style program. Unlike a scene, a sketch leaves no lingering questions of "What will happen next?" at its completion. When a sketch is over, something completely different will probably follow (a song, an entirely different time and space, etc.).

In addition, sketches are overtly performative. Hollywood scenes require performers, to be sure, but these scenes are constructed to downplay the fact that they are performed for an audience. The classical film strives to convince us (at least temporarily) that these scenes take place in a realistic facsimile of our world, and that we are watching people/characters, not actors. Only extremely rarely does a film actor turn to acknowledge the presence of an audience watching in the dark (and many of those, such as Groucho Marx, come from a background in sketch comedy). A sketch, however, tends to present a broader style of acting than the realistic style that dominates mainstream film. They are more likely to acknowledge that they are performing for an audience, not pretending to be "real" people.

The structure of sketches in an evening's entertainment, in which members of the troupe play a variety of parts, often foregrounds this performative unrealism. In a Python sketch the actor playing a game show host in one sketch might be a Minister of Silly Walks in another; little effort is expended to conceal the fact that this is the same person. In a classical film it is expected that different characters will be portrayed by different actors, continuing the realistic facade. If a classical film recycled its actors in different parts, it would either be a display of bravura acting virtuosity (e.g., Alec Guinness) or an admission of an extremely low budget. Following the tradition of sketch comedy, *Grail* flaunts the fact that its players play multiple roles. John Cleese plays both Sir Lancelot and the rude Frenchman who taunts Arthur's knights. Eric Idle is Sir Robin, Roger the Shrubber, Brother Mayor, and one of the heads of the Three-Headed Knight. These multiple performances help signal that we are watching sketches, not merely scenes.

Since sketches are not as tightly bound into a narrative structure as are traditional scenes, they tend to be freer to pursue "extraneous" material. Without the scene's requirement that it must further the protagonist's quest for a goal, the sketch can be derailed by what would otherwise be "tangential" subjects. In *Grail*, for example, several conversations (at a castle wall, a witch trial, and the Bridge of Death) unexpectedly veer into a discussion of the relative airspeeds of African and European swallows. A king's simple attempt to gain information from a peasant can change into an intellectual discussion of the rights to sovereignty (of the Lady in the Lake story, the peasant says, "Strange women lying in ponds distributing swords is no basis for a system of government!").

The sketch structure of *Monty Python's Flying Circus, And Now for Something Completely Different*, and *Grail* makes it easier for this material to be lifted out of its original context and recirculated in others. Since sketches are relatively freestanding, they can be used without explanatory frames. The performative quality of Monty Python sketches has helped open up a new channel for circulating this material: in the everyday interactions of Monty Python fans. Before *Quest* recycled material from the *Grail* film, this material was primarily recirculated among Python fans. In order to better understand how *Quest* recycles Python quotes, we first need to examine (at some length) how fan culture recirculates "quotable" sections from Python sketches. The quotable structure of *Grail*, as we shall see, makes it easier for both fans and CD-ROM designers to resituate this material onto a CD-ROM or into everyday life.

Monty Python's cult following does not depend solely on the texts themselves; instead, fans interweave sketch material into their own daily lives. "Jokes, characters, and catch-phrases became the secret signs of members of the cult, and the Pythons found their jokes replayed to them," Robert Hewison (1981, 8) has noted. *Grail*'s status as a mainstay of midnight movie screenings has enabled countless fans to memorize crucial lines and often entire sketches from the film, storing them away for further use in everyday situations.

Here the performative nature of the material plays an important part in its recirculation. Part of the fun (particularly for American audiences) is nailing the "outrageous" accents of the Python troupe members. It is not simply enough to repeat lines from the film. Fans try to reperform the lines, mustering whatever impersonative vocal skills they have. My undergraduate roommate was fond of Sir Lancelot's reply after he massacred the wedding party (Crowd member: "But you killed the bride!" Lancelot: "Sorry. Terribly sorry about that"). Merely saying the words does not mark the phrase as a Pythonesque quote; one has to imitate Cleese to bring off the rejoinder. Just as Python members slip back and forth among roles (from Sir Lancelot to Tim the Enchanter), Python fans slip back and forth among different performed voices (from their standard voices to Denis the Peasant to the King of Swamp Castle).

Henry Jenkins (1992) has shown how fan practices often convert consumers of media into producers of media, making stories, videos, and artworks based on their favorite TV texts. Python fans become producers as well, but they tend to do so with a slightly different emphasis than the *Star Trek* and science fiction fans Jenkins studied. Python fans produce through performing on the everyday stage. While *Star Trek* fans may latch onto the occasional catchphrase ("He's dead, Jim"), the sketch structure with its emphasis on verbal play encourages a much greater attention to language. Python fans tend to be performers who resituate familiar material into new everyday contexts.

These catchphrases, liberated from their original context, can be applied to a variety of everyday situations. Once you start using a phrase such as Lancelot's "Sorry. Terribly sorry," you begin to realize how frequently situations call for such an inadequate apology. Bits of *Grail* dialogue become quite useful in one's personal rhetorical bag of tricks. I am quite fond of the Frenchman's taunt, "Your mother was a hamster, and your father smells of elderberries." It's astonishing how often one needs a good, all-purpose insult like this one.

The tangential quality of Python sketches contributes significantly to the adaptability of these lines. Unlike more traditional Hollywood dialogue, these lines do not seem particularly bound to a specific narrative situation. In *Grail*, discussions of relative swallow velocities or the desire for a shrubbery have little bearing on whether or not Arthur and his knights find the Grail. Lengthy instructions to a particularly dense castle guard or Socratic dialogue concerning how to test for witches have bearing on the progress of the individual sketch, but not on the outcome of the whole film. This material seems just as well- (or ill-)fitted to my own everyday interactions with twentieth-century bureaucrats as it does to this medieval quest narrative. Because this dialogue is so tenuously tied to the subject at hand, it is easily resituated into fans' everyday lives.

Fans are not limited by the structure of texts, of course. They can choose whatever lines they want to circulate as catchphrases among themselves. But the point is that *Grail*'s structure of freestanding, tangential, performative sketches makes it easier for fans to incorporate bits of the film into their lives. The sketch structure encourages this more than typical Hollywood construction.

The *Quest* CD-ROM reuses *Grail* material in a manner somewhat between film watching and everyday quote performances. Several of *Quest*'s sites re-present clips from the film in a more overtly interactive fashion. One must click on the various characters to make them speak familiar lines from the plague sketch (in which a body collector is convinced to dispose of an old man's body, even though he's "not dead yet"), the witch trial scene, or the French taunting episode. This particular form of interaction gives the pleasure of anticipating and experiencing a familiar sequence of lines, just as rewatching the film does. After a member of the lynch mob testifies against the witch ("She turned me into a newt!"), the clarifying addendum to his statement ("I got better") is only a click away on the CD-ROM. One can interrupt the familiar sequence by clicking on another object, but one cannot radically reshuffle the lines into an entirely new order. Although this form of interaction feels quite limited, it does give some sense of the performative quality of using Python quotes in everyday life. You control the timing of these lines' delivery. Without your clicks, the scene will not progress. Instead of participating alongside the diegesis in the film by saying the lines, one participates in the CD-ROM's diegesis.[5] Like the Python fan who uses the lines in everyday situations, you control when the lines are spoken.

The designers of *Quest*, like Python fans, recognize that the *Grail* catchphrases can be used in a variety of situations other than their original contexts, and so they do not limit these quotes to their familiar settings. Instead, the quotes are scattered throughout the CD-ROM, demonstrating further how remarkably adaptable the lines are to different contexts. In the film, Arthur's stately attempt to introduce himself as king is met with the skeptical retort, "Pull the other one!" This line seems even more appropriate when spoken by a skeleton in the Black Knight's tent in *Quest*. Arthur's lines "We must watch and pray" and "You're a loony!" become useful in a variety of situations. "Run away!" and a dispirited choral "Yay" from the film become repeated structuring units in the CD-ROM. In linking quotes with different contexts in the CD-ROM, the designers take their cue from Python fans, who transplant quotes onto other situations. In so doing, *Quest* makes surprising connections between new and old material.

Henry Jenkins (1992), following Michel de Certeau, has discussed how fans "poach" from popular culture, appropriating imagery that they use for their own purposes in their daily lives (24–36, 44–49). Certainly Python fans who use bits of sketches in their everyday

Fig. 3.1. "She turned me into a newt!" A click on the witch's trial scene yields quotable lines from the film.

conversation are "textual poachers" in this sense. The processes by which fans choose to poach one thing and not another are complex and still not very well understood, but this choice obviously must involve factors both in the text and in the individual. All texts are not equally susceptible to this process of poaching, although the text itself cannot exclusively determine whether or not it will be poached. It would have been difficult in 1975 to predict that a medieval parody/ satire film would instigate decades of repartee among a devoted cult following.

With the benefit of hindsight we can see that certain distinctive characteristics of *Monty Python and the Holy Grail* made it more likely to be poached. The picaresque structure of Arthurian tales dovetailed with the sketch structure carried over from the *Flying Circus* to create a series of relatively self-enclosed sketches that did not rely on a strongly linear sequence.[6] The tangential quality of the dialogue (which frequently wanders from the ostensible initial direction of the scene) further casts loose from the moorings of the film's overall plot.

These structural characteristics made it easier to treat the film as a series of "quotable" moments. Furthermore, the performative quality of these moments gave additional pleasures to the fans who integrated them into their lives.

I argue that certain pop culture texts are more "quotable" and therefore more likely to be poached. The term "quotable" is intended to indicate that both images (Marilyn Monroe's dress being blown upward) and dialogue may be removed from their original contexts and integrated into fans' everyday lives. What determines whether a film is quotable? An exhaustive list of characteristics is probably not possible, given the creativity of fan practices. However, from the preceding examination of *Grail* we can list some qualities that make a scene more quotable: the ability of the scene to stand alone; an emphasis on overtly marked performance; a tendency toward tangents.[7]

This history of fan participation through the recycling of quotes helped inspire the creation of *Quest*. 7th Level cofounder George Grayson came up with the idea for the disc after noticing that computer programmers are frequently big Python fans (Smith, 1994).[8] These same factors that make *Grail* quotable within the broader culture also help to facilitate its translation to CD-ROM. Because fans demonstrated how discrete chunks of the film could be used in a broad range of situations, *Grail* became an ideal candidate for the nonlinear medium of CD-ROMs. Given the participative nature of these sketches, the designers realized that this source material could be adapted to a more "participatory" medium.

What does this mean for translating films to the new medium of CD-ROM? The first thing to note is that the classical Hollywood narration is designed for spectator immersion, not quotability. The structure is set up to propel a viewer forward through the narrative, not to set apart certain sections for viewer poaching. The key factors in understanding *Grail*'s translation to the *Quest* CD-ROM are those characteristics that distinguish *Grail* from most Hollywood films. This would seem to reassert the difficulty of adapting most Hollywood texts into CD-ROMs.

However, this is a statement of a trend, not a totalizing judgment on the prospects of film/CD-ROM adaptations. We need to look more closely at instances of quotability in the classical Hollywood system. For instance, recent action/adventure films have made a conscious effort to contain at least one catchphrase ("Hasta la vista, baby," "Go ahead, make my day") that producers hope will circulate among the population at large, thus promoting the film. These moments are consciously intended to be quotable and marketable, and investigating the structure of these moments may help us understand quotability better.[9]

Although the classical narrational system does not promote quotability, the fact is that fans have always found a way to poach elements from it. As fans continually discover new poaching strategies, we should look to them for a better understanding of what makes a text particularly quotable. Fans integrating film quotes into their everyday lives could be considered a testing ground for multimedia adaptations of movies.

And Now for Something Completely Digital

Grail's quotable structure makes it easier to import discrete elements (particularly dialogue and film clips) into the CD-ROM adaptation. But *Quest* is not composed solely of elements from the film rearranged into new contexts. 7th Level designers also translated several Pythonesque strategies into their CD-ROM equivalents, and these translations help extend *Quest* beyond the influence of the original material. I will examine how several Python strategies (interruptibility, stop-motion animation, parody, and satire) are reconfigured into CD-ROM equivalents.

Although Monty Python's comedy is clearly within the time-honored tradition of sketch comedy, it also departed from that tradition in significant ways. Graham Chapman noted that the *Flying Circus* differed "from most TV sketch shows by ignoring the conventions established over the years—that sketches must have a beginning, middle, and end, and a punch line, above all. They must also be interspersed with songs and dances—that was always the tradition, a holdover from stage variety or stage revue" (Johnson, 1989, 5). Python sketches did not always have the structural wholeness of traditional sketch comedy ("a beginning, middle, and end"). With little warning the comic proceedings might be interrupted by the figure of John Cleese, sitting behind a desk situated in an incongruous setting (such as a beach or a field), announcing, "And now for something completely different." In other instances sketches might be rudely interrupted by a sixteen-ton weight crushing a particularly annoying character. Or a band of robed priests might burst into a sketch, bringing with them the Spanish Inquisition.

Monty Python presented sketches, but it also violated the conventions of British comedy sketch structure by frequent interruptions for "something completely different." The troupe members were attempting to supplant the orderly structure of the TV comedy sketch with something nearer "stream of consciousness, like a hilarious bad dream" (Hewison, 1981, 8). According to this dream logic, it made just as much sense to interrupt a sketch with a short, irreverent animated sequence as it did to complete the sketch. Closing credits need not be seen only at the end of the program; instead, they could appear at the middle or the beginning of the show. Interruptions of the normal sketch structure became the norm in *Flying Circus.*

Grail continues this interruption strategy and the Python preoccupation with misplaced titles. The film's narrative is interrupted by a misleading intermission title, animated monks in robes diving off a diving board, and an animated multi-eyed dragon (which in turn is interrupted by the death of the animator). The opening credits halt (after displaying a curious preoccupation with Sweden) while the credit writers are sacked, and then the credits continue (at great expense and with a curious preoccupation with llamas). The primary medieval story is intercut with scenes of a present-day historian authoritatively discussing Arthur's quest. After the historian's throat is cut by a marauding knight from the other plot line, police investigate the death, and the film occasionally (and briefly) cuts back to the investigation before returning to something different from the Middle Ages.

As discussed earlier, *Grail*'s primary narrative is already segmented into sketches. The interruptions further loosen the moorings that bind this material to a linear narrative progression. The interruptible style of Python humor made *Grail* more easily adaptable to a nonlinear medium. Since *Grail*'s diegesis could obviously be interrupted by animated material, sequences from the present day, or nonsensical credits, the designers of *Quest* could easily interrupt the material with games and pop-up figures without doing violence to the original intention.

In fact, the *Quest* CD-ROM seems to carry Python interruptibility further. The intertitles from the film are repeated with numerous new variations ("Did you know your fly's open?" "Fooled you"). As noted earlier, clips from the film are re-presented as interactive scenes that frequently halt, requiring our clicks to start the scene again. The CD-ROM opens up many possibilities for players themselves to create interruptions of their own, since players can choose to click on various objects or to visit other locations in the middle of the scene.

Because the CD-ROM offers both prescripted interruptions (e.g., the intertitles) and the opportunity for player-initiated interruptions (through clicking), it extends the innovative interruptible style of *Monty Python's Flying Circus*. The interruptions that helped make Python sketch comedy distinctive become a structuring principle for the construction of a nonlinear CD-ROM.

One of the most distinctive ways to interrupt Monty Python sketch material (both film and television) is to interject one of Terry Gilliam's stop-motion animation sequences, which he prepared without the troupe's collaboration to be inserted into the sequence of sketches. Gilliam made no pretense to be doing smoothly professional animation. Instead, the moving drawings and cutouts were intentionally jerky; they seemed as irreverent toward the standards of animation as they were toward the standards of propriety. The jerky motion reminds us that animation is based on a series of still pictures, not on an illusory continuous motion. Such animation gives these sequences a "homemade" quality that fits the low-budget aesthetic of the *Flying Circus*. This overt admission of budget constraints helps situate these sequences as lower-class puncturings of upper-class pretensions. The jerky animation tells us that Python's satire is from the less-well-funded margins, not from the center.

The *Quest* CD-ROM continues the Gilliam tradition of jerky animation and extends it throughout the Python text. In the film, stop-motion sequences of monks in hooded garb diving off a diving board or a cutout stained-glass Jesus waving a blessing are exceptions to the smooth motion of the rest of the film. In *Quest* the entire CD-ROM uses jerky stop-motion animation. Cutout figures of Arthur and his knights reenact scenes from the film, and these figures also pop up from the margins to give commentary (e.g., Arthur pokes his head in from offscreen to say, "We must watch and pray"). In each of these situations, the cutout versions of Python members in medieval costume speak with their mouths clearly not in synch with the dialogue.

Instead, their mouths open and close as if hinged, resembling Gilliam's animated style more than the smooth motion of film.

Here the 7th Level designers have found an aesthetic justification for one of the CD-ROM medium's limitations. Because smooth computer animation requires considerably speedy processing and sizable memory, many CD-ROMs have rather poor animation quality. The *Quest* CD-ROM takes this limitation and turns it into an aesthetic asset. The jerky motion of the cutout figures seems to fit the animation style familiar to Python fans. Now the Python troupe members move according to the same rules governing Gilliam's creations. Just as Gilliam acknowledged the limitations of his low-budget stop-motion animation, the *Quest* designers acknowledge the shortcomings of the CD-ROM medium.

I have argued elsewhere (Smith, forthcoming) that when CD-ROM designers provide a diegetic rationale for the limitations of the CD-ROM medium, it helps hide the "computerness" of the program in a way that is elegant and involving. *Myst*, for example, tells the story of several islands whose populations have been wiped out; thus the CD-ROM's construction as a series of still images is justified (since nothing on these islands is left alive to move). *Myst* avoids awkward CD-ROM animation by creating a rationale for avoiding movement completely. This strategy (along with *Myst*'s lack of an overt interface, such as a menu) encourages us to get caught up in the world of the story, not to contemplate the frustrations of the CD-ROM medium. *Myst* hides the "computerness" of the disc, which helps us concentrate primarily on the densely detailed worlds.

Quest relies on an entirely different strategy to encourage viewer involvement. Instead of concealing the "computerness" of the disc, it flaunts the fact that this is a CD-ROM. The intentionally jerky animation constantly reminds us of the medium's annoying limitations, but by incorporating this awkward motion into a Pythonesque aesthetic, the *Quest* CD-ROM makes available a different sort of pleasure. By emphasizing its computerness, *Quest* encourages the player to laugh at the medium's characteristics. While *Myst*'s appeal depends on its seemingly invisible interface, *Quest*'s pleasures depend on an awareness of how clunky a CD-ROM interface is. Both strategies encourage player involvement, but they do so in different ways. A disc that deemphasizes the computerized nature of the text tends to promote immersion in the story world. A disc that flaunts its computerness opens up the possibility of parodying the medium itself.

Parody and Satire: Wasted Time

Parody, as Linda Hutcheon (1985) discusses it, is repetition of elements from other works of art with the addition of critical ironic distance.[10] It is imitation of formal properties with ironic intent. From its earliest *Flying Circus* seasons, Monty Python has parodied the formal properties of the television medium itself. We have already mentioned Python's tendency to misplace titles; they also would break other unquestioned rules of station continuity by showing station identification from the other channel. Python sketches used television genre forms with an absurd twist (such as a game show in which Mao, Lenin, Marx, and Che Guevara eagerly answer English football trivia questions), which pointed up the ridiculousness of the original form. Pythonesque humor is not merely silly bits; it reuses existing forms, adding what Hutcheon would call an ironic distance.

Clearly *Grail* is a parody in Hutcheon's sense, a send-up of the form and content of the Arthurian tales. By taking Sir Lancelot's gung-ho attitude to an extreme (slaughtering people first, asking questions later), by placing Sir Galahad the Chaste in a situation promising orgiastic sex, and by inventing the cowardly Sir Robin, the film skewers the notion of knighthood as chivalrous, pure, and brave. When the fiercest obstacle in Arthur's path is a man-killing rabbit, *Grail* parodies the structure of innumerable dragon-slaying tales. For those familiar with legends of the Grail, the film presents irreverent surprise after surprise.[11]

While its send-up of medieval imagery is quite funny, a medieval parody is a fairly arcane idea for a mass audience. *Grail*'s parody extends beyond the specific form and content of Arthurian tales to make fun of film itself. *Grail* continues the Python strategy of lampooning the medium it is using, violating its unquestioned assumptions just as the *Flying Circus* did with television. The film announces its parodic intent from its initial moments, in which we dimly see Arthur and his servant riding through the countryside accompanied by the sound of horses' hooves. After a while we see Arthur and his servant top a hill, and only then do we see that the clopping sound is not made by horses but by two coconuts that the servant is banging together. By placing the sound effect process (normally hidden behind the scenes) in the foreground, the film exposes its own low-budget approach to parody. *Grail* even takes potshots at the musical's tendency to provide orchestral background at a moment's notice. Herbert, the "damsel" in distress, has a tendency to burst into musical production numbers, but his father keeps squelching the unseen orchestra. This overt acknowledgment of the conventions of filmmaking provides parodic pleasures in addition to the send-up of Arthurian imagery.

Similarly, *Quest* does not settle for echoing *Grail*'s medieval parody. It also spoofs the medium of the computer game itself. *Quest* presents dead-on parodies of well-known individual computer games and other familiar computer applications. Included in the CD-ROM is the "Knights in Kombat" game, which duplicates the imagery from the violent *Mortal Kombat* (from its opening music to the final command, "Finish him!"). "Knights in Kombat" resituates this violent game onto the context of Arthur's limb-hacking battle with the Black Knight. In addition, the Black Knight's journal is a cross between Atrus's books in *Myst* and a sappy adolescent diary. After being commanded by the knights who say "Ni!" to find a shrubbery, one can access the "Dial Intershrub" program, which mirrors a World Wide Web browser. Players can then access a not-very-helpful help function that speaks fake Swedish ("Fur cursor to de left mooven, mooven de moose to de left"), find out "What's Ni," or simulate dialing in to purchase shrubbery "online." *Quest* recognizes that a Python program should parody other programs, whether computer or television.

One of the most distinctive things about this CD-ROM is that it consistently lampoons basic assumptions about game play. Although heralded as a "new medium," CD-ROM games have already established a set of norms, and *Quest* consistently makes us conscious of the medium's conventionality. It pokes fun at shooting indiscriminately at objects, a basic organizing principle for countless games. Instead of shooting at menacing monsters as in *Doom*, *Quest* asks us to chase down and shoot ridiculous objects such as flowers. *Quest* takes the conventional computer prompt "Do you want to play again from the same place?" rather literally in the encounter with the lethal white rabbit. After the bunny savagely dismembers your knights, you can choose to play again, an option that loops you back to *exactly* the same

Fig. 3.2. Monty Python and the Quest for the Holy Grail parodies the *Mortal Kombat* game in a bloody struggle between Arthur and the Black Knight.

place, where your knights can be massacred again and again (until you discover that you really want to "play again from a slightly different place"). It makes fun of the now-familiar process of answering "just a few questions" when registering a new CD-ROM. *Quest* presents you with a barrage of 125 questions, some of them familiar to *Grail* fans (involving velocities of swallows, for instance), some of them a bit too personal ("Are you wearing underwear? Do they fit snugly?"), and some simply inane ("Bing tiddle tiddle bang?"). The player cannot skip a single question in this interminable registration process if he/she is to complete the game.

Beginning with the product registration, playing a CD-ROM combines pleasure and frustration in alternate measures, and *Quest* constantly points out this frustration. Throughout the game *Quest* attacks the inanity and tedium of the central process of playing a CD-ROM: clicking. Clicking the mouse is the key to unlocking the pleasures of most non-joystick games, but such clicking results in enormous frustration. Players click on objects that don't perform in the ways they hoped, leading the player to click on every possible object on the screen in an attempt to unravel the secrets of the virtual space.

Clicking in *Quest* flaunts how pointless such clicking can be. Mouse clicks elicit sound bites that can be overtly unsatisfying ("You've done it! You've clicked on the spot!"), misleading ("This is where you should've clicked in the last scene"), or scolding ("Don't click here. This is my spot. I found it first"). *Quest* spoofs the interminability of game clicking with such phrases as "Click here 500,000 more times and you get a surprise!" and "Isn't your finger getting sore yet?" At the same time, however, *Quest* overtly requires us to engage in the same frustrating, repetitive clicking it parodies. Several objects require multiple clicks before they will provide game-crucial information. To get to see the "gratuitous act of violence" promised in the plague scene, one must click repeatedly and determinedly. By simultaneously scoffing at clicking and forcing us into repetitive clicking situations, *Quest* reminds us of the frustrations that are part of computer game play. This CD-ROM encourages us to confront the ridiculousness of a medium that forces all interactions to take the form of a single button press.

In this manner, *Quest* veers more toward satire and away from pure parody. Linda Hutcheon usefully differentiates the two terms, which are often confused. She uses parody to refer to the ironic recirculation of material from other texts. The target of parody is other texts, particularly their formal qualities. Satire, on the other hand, points outward toward the outside world beyond formal texts to the foibles and follies of humankind. Its target is social and "extramural," not "intramural" (between texts). According to Hutcheon, part of the reason satire and parody are often confused is that they often are used together. A satirist who wants to send up a particular social convention will often parody a textual form to gain attention. For instance, a Monty Python sketch satirizing upper-class pretensions might choose to do so by parodying television sports competition, creating the race for "Upper-Class Twit of the Year."

Monty Python's work has always had a strong emphasis on satire, with a particular focus on British aristocrats, bureaucrats, and petit bourgeoisie. *Flying Circus* featured an array of bizarre shopkeepers: one who alternates between being helpful and outrageously insulting; a pet shop clerk who refuses to take back a dead parrot; and a mortician who proposes that bodies be cooked and eaten. John Cleese satirized useless British ministries in his Ministry of Silly Walks sketch. Clerics were also subject to ridicule, as in the "Dirty Vicar" sketch (where a vicar uncontrollably grabs women's bosoms). Python made fun of national stereotypes, as in the sketch about an Australian philosophy department where they propose that all faculty members be named Bruce. They also bent gender stereotypes with singing lumberjacks who "put on women's clothing and hang around in bars" and "Hell's Grannies," a violent gang of delinquent old ladies. Often through simple inversion, Monty Python would put stock figures into situations where they would normally never appear. They placed people who would never meet together in the same sketch. Removed from their usual context, these characters' pretensions and vices become obvious targets for satire.

Grail is the single Python film that tends almost entirely toward parody and neglects the Python satiric tradition. Much of the fun in the film comes from the play with Arthurian myth, the conventions of film, verbal tangents, and blatant anachronisms, not from satire of real-life figures. Occasionally the film does veer toward real-life subjects (e.g., the confrontation between the decidedly modern politics of an anarcho-syndicalist peasant and the monarchic politics of a king), but it avoids the class-based satire that characterized much of the Python work in *Flying Circus*. 7th Level's *Quest*, following the lead of the original film, tends toward parody, but it also reactivates the Pythonesque tradition of satire in an important way.

Quest enters into satiric territory by emphasizing the notion of wasted time. This concept, according to Jeremy Rifkin, was made possible by the invention of the clock:

> Slowly, the bourgeoisie began to advance the idea of securing the future through the proper husbanding of time. The clock became the instrument to hoard and mete out time. The clock's introduction into the economic life of Europe led to the idea that time could even be bought and sold. . . . If time could be bought and sold in units, it could also be accumulated or depleted. In the new clock culture, time and money became interchangeable and exchangeable. (161)

An economics of time developed, which continues in force to this day. If one packs more activity into a given period, one is "saving" time. There are values in this economy: if one engages in a worthwhile activity, one "spends" time, but activities that are considered worthless "waste" time. The spending and wasting of time is measured against the future. By purposeful and efficient husbanding of time in the present, one gains potential time to be "spent" in the future. Spending time is an investment; afterwards, one gains a certain return (by spending time with my child, I help our relationship to grow). Wasting time produces no such return. Spending time involves a sense of conscious choice to do an activity, while wasting time evokes a sense of aimlessness and lack of agency.

Much of popular entertainment is caught between these notions of spending and wasting time. People acknowledge that they watch television as a way of satisfying their need to unwind after a hard day's work, which would indicate that they are spending time (actively choosing an activity that serves a need). However, watching television is widely considered a waste of time, characterized by aimless wandering through channels with little overt return on people's time. Discussions on the value of popular entertainment often center on these discrepant perspectives. Pop culture either serves no overt purpose (and therefore wastes time) or serves a hidden purpose (which can be worthwhile, such as "letting off steam" or relaxing).

This dichotomy in the economics of time structures the way we think about new media, particularly computerized media. Time economics has increased in importance as the pace of society increased, and the computer is on the cutting edge of this speed-up process. Personal computers are now capable of multitasking, carrying out more than one process at the same time to provide maximum efficiency. The aim is decreasing the amount of time a user must wait for the computer to finish processing. Increasingly speedy processing chips and superfast modems offer the promise of quicker loading of programs and downloading of files. The computer has become a symbol of relentlessly efficient, rapid use of time.

The ever increasing expectation of computer speed often collides with the limitations of actual hardware and software. Lured by the promise of more efficiently spending our time, a computer user becomes increasingly aware of wasted time waiting for image files to download from the Internet or new programs to complete installation. As time economics speeds up, the computer that seemed fast to me a year ago now feels agonizingly slow. Increasingly, the experience of using a computer increases one's time frustration. The user, once amazed at what the computer can do, is now frustrated that it can't do things faster. We internalize a computer pace that quickly escalates to outstrip the computer itself. Although the computer seems a device designed for ultimate time efficiency, actually using computers emphasizes the many small wastes of time.

While computers can magnify our perception of the frustrations of wasted time, they also can provide increased opportunities for pleasantly wasting time. Although the personal computer was initially marketed for its efficient, practical functions, the game capabilities of these computers became the key to their widespread purchase. Large numbers of people bought these purposeful devices only when it became clear that one could enjoyably waste time on them. Attuned to this tension between wasting and spending time, many computer games reiterated the economics of time, as Rifkin noted: "The ultimate goal in most computer games is to secure more time. Every aspiring video game player dreams of the perfect game, the game that goes on forever. Time, in computer video games, is a foil, a resource, and prize all wrapped up as one" (25). Computer games lure with the promise of a return on the time you've spent playing them. The lure is more playing time, which gives the player a sense of accomplishment. By rewarding the player's efforts with "time," games mimic the economics of time. However, the sense of accomplishment at gaining more playing time does not necessarily help allay the sense that game play overall is a waste of time. Just because you've been rewarded with more playing time doesn't mean that playing isn't a waste in the first place.

CD-ROM games are played in the context of this network of social meanings and values. These games, like television and other popular media before them, are often caught between the concepts of spending and wasting time. People sit down to play *Myst* or *Doom* for a few minutes and emerge hours later, astounded at the time that has passed and often guilty over the time they have wasted. But the very fact that these games are played on a computer accentuates the spending/wasting dichotomy. Playing a game (which, by definition, has no practical purpose) on the very machine that is supposed to bring time efficiency to one's life simply increases the tension between spending and wasting time.

Quest is unusual in the way it overtly, wholeheartedly, unashamedly announces that it is a waste of time. It flaunts its own uselessness. Users determinedly clicking in their quest for the Holy Grail hear comments such as "To waste more time, click here again." A player leafing through the pages of the Black Knight's journal hears this comment: "I could keep turning pages all day" (which might easily have been said by a *Myst* player reading Atrus's journal). Sound bites such as these begin to move into satire's territory. Not satisfied merely to parody clicking, *Quest* pokes fun at modern time pressures, which few other CD-ROMs do.

Immersive CD-ROMs such as *Doom* do not encourage the player to consider that he/she is wasting time. Only after one finishes a playing session is one struck by how much time has passed (wasted). This awareness is not triggered when one plays the CD-ROM but when one makes the transition from game time to real time. But *Quest* contains numerous reminders that while you are playing, you are wasting your time. Not coincidentally, Monty Python's first CD-ROM (composed mostly of famous sketches from *Flying Circus*) was entitled *Monty Python's Complete Waste of Time*. From the titles of their CD-ROMs to the senseless tasks they set for the player, Python's CD-ROMs flaunt the knowledge that other CD-ROMs hide (but which many of us secretly share): CD-ROM game play is a waste of time.

And why not? By joyously announcing its uselessness, *Quest* encourages us to confront the social stigma of time wasting. In a society that increasingly asks more and more of our time, a little wastefulness is illicit fun. It is a jab at the ever present voices that say, "Faster! Faster! More! More!" Part of the enjoyment of wasting time is the knowledge that one really should be doing something more productive. *Quest* doesn't remind us that we're wasting our time as a means of inducing guilt. Instead it emphasizes its uselessness to point out how silly the social pressures toward time efficiency are.

The computer game is an important factor in the time wars (as Rifkin calls them). Employees forced to do less-than-involving work on a computer can sneak in a little game playing (some games provide a built-in function to display a fake word processing screen to conceal the game from a passing supervisor). By snatching a bit of employer time to gain some personal pleasure (a strategy de Certeau calls the *perruque* [1984, 24–28]), the employee fights back against the time pressures. Taking back time by using the computer, the same resource your employer provided for practical purposes, accentuates the illicit pleasure of wasting time.

Quest is no different from many other computer games in its time-wasting capacity. Much of the pleasure in *Quest*, as in many CD-ROMs, comes from clicking on objects and watching them do silly things (such as clicking on the witch to watch a long lizard tongue come out of her mouth). The difference is that *Quest* consistently reminds you that you are wasting your time, thus satirizing the logic of efficiency that interpenetrates the world of computers.

The early academic pronouncements on hypertexts and hypermedia emphasized these new media's capacities to destabilize linear logic and embody more poststructural ways of thinking.[12] As these forms have evolved into multimedia, they have become conventionalized, as all new media do. Within the conventions that have emerged through commercial development of CD-ROMs, however, there is little of this radical critique of logic. In fact, many CD-ROM games seem to reiterate the primacy of logic by emphasizing intricate puzzle solving and systematic exploration of the virtual space. Such programs reinforce the logic of the computer instead of restructuring it.

While staying within the context of commercial multimedia, the 7th Level designers have found a pleasurable way to break down this logic. The potential for this playful illogic came not solely from the structure of the CD-ROM medium but through the adaptation of Pythonesque strategies to the new medium. It is not merely the parody/satire of Monty Python that provides the illogic, nor does the nonlinear structure of multimedia alone result in a reconfigured logic, as early critics promised. It is the combination of parody/satire and nonlinear media's capabilities that gives *Quest* its critique. Satire and parody encourage a reflexive stance that, in conjunction with multimedia's interactivity, points us outward to other texts and to the outside world. As developers continue to adapt and recycle existing material into commercial CD-ROMs, it is not enough merely to "lift" material from the source and hope that multimedia's capabilities will "add" interactivity. As *Monty Python and the Quest for the Holy Grail* demonstrates, the key may be first to find the distinctive strategies in the text and then to encourage those strategies and the CD-ROM medium to transform each other. This is the Holy Grail of multimedia: a shiny new medium transformed and energized by interactions with the old.

NOTES

1. Hollywood films that have been adapted to computer/video games include *Batman Returns, Predator, Wayne's World, Fantasia, Rambo 3, Beauty and the Beast, Home Alone 2, The Last Action Hero,* and many others.

2. The quest is one of the primary organizing principles in the history of computer games. Following *Adventure,* many games are structured around wandering through a treacherous, unknown space trying to collect important objects. The Grail quest is the prototype of such narratives. This deep similarity between computer game structure and the source material for *Grail* aids in the translation of this material to CD-ROM.

3. It is interesting that this interactive CD-ROM rewards you not by giving you more interactive experiences but by giving you film clips, which we can only watch without interruption or interaction. The final reward is itself a film clip. Other CD-ROM games also present film (or rather, QuickTime) excerpts as payoffs. For instance, when you finally break through to another age in *Myst,* you are shown an aerial QuickTime flyby of the new age's landscape, and this reward is quite satisfying and involving, in spite of its reliance on "old-fashioned" film/video images. Perhaps a switch to another medium within a multimedia text (from game to film, for instance) is enough to signal a "bonus" for the player. However, it may be more difficult to use a film clip as a final reward (as both *Quest* and *Myst* do), since we expect a more strongly marked payoff from moments of closure.

4. Hollywood has occasionally produced multiple story films using discretely framed segments (e.g., *Dead of Night, New York Stories*). Such films do not radically restructure classical Hollywood storytelling norms, since the individual tales tend to follow these conventions. However, the rarity of such films reminds us that they are an exception to the overall structure of feature film narrative.

5. The quality of this different form of participation has moral consequences. In the plague scene a body collector is finally persuaded to clonk an old man on the head, in spite of his protest that he's "not dead yet." I have laughed many times at this scene in the film with few moral twinges. It was quite another thing, however, when the fatal clonking required my click to silence the old man's cries ("I feel happy!"). Similarly, watching on film as King Arthur hacks away all four of the Black Knight's limbs provides a sort of moral distance that helps create the humor. When you are involved in a game situation in which your goal is to hack away his limbs, this emphasizes your own participation in the comic sadism. This is further accentuated because the Black Knight inexplicably scampers away faster and faster as you hack away more of his limbs, which puts you in the position of clicking furiously as you wield a tiny sword and chase a torso with no legs. Similar

arguments could be made for the fast and furious participation required in the games included on *Quest*, including "Spank the Virgin" (where one raises red whelps on the lily-white bottoms) and "Burn the Witch" (a memory game that rewards you for setting afire the correct sequence of witches tied to the stake). I do not wish to sound puritanical about the violence in *Quest*. On the contrary, the ridiculousness of the violent tasks and the requirement of my participation set up a space for me to think about my own moral culpability.

6. Because the sketches in *Grail* are not particularly interdependent, it is difficult to remember the order in which they occur in the film. Is Sir Robin's tale before or after Lancelot's? Does Arthur's confrontation with Denis the anarcho-syndicalist peasant occur before or after the witch trial? Because the film does not have a strong chain of causality binding the episodes together, I find that the only way I can mentally reconstruct its sequence (without revisiting the text) is to recall which players are involved in the various scenes. Since the Wooden Rabbit scene involves Sir Bedevere, it must take place after the witch trial (where we meet Bedevere).

7. Given this brief, incomplete list, I note how well it describes some of the film texts that my undergraduate students quote most often. *Grease*, with its structure of self-enclosed musical numbers and mannered musical performances, frequently works its way into my students' interactions, as do the more talky, tangent-oriented John Hughes films such as *The Breakfast Club*. Certainly there are thematic factors involving adolescent confusion and pressure leading my students toward these two films. But there are also structural factors that make certain films such as *Grease* and *Grail* more quotable for fans.

8. In an article reporting on multimedia products (including *Monty Python's Complete Waste of Time*) premiered at Comdex (the most important computer trade show), David English (1994) reported that the "highlight of the evening" was when the crowd of computer professionals sang Python's "Lumberjack Song" along with the CD-ROM.

9. The action/adventure film may also show how films may alternate between spectacle and narrative. The advancing plot in an action/adventure film frequently comes to a halt to present a moment of spectacle (a car chase, a city exploding under alien attack). These relatively self-enclosed moments may be more quotable and therefore more able to be included in CD-ROM adaptations. This harks back to the concept of "attractions" pioneered by Sergei Eisenstein and further advanced by Tom Gunning (1986). An attraction (like a circus sideshow or a sexy picture) appeals to an audience without requiring narrative setup. Gunning has used this term to describe early cinema, which tended to present such attractions without much narrative justification. To the extent that our blockbuster cinema is returning to a cinema of attractions, this may help designers adapt more quotable films to nonlinear media.

10. Hutcheon differentiates parody from other related forms. Parody is more specific than mere intertextuality, since intertextual citation can be made without ironic intent. Allusion emphasizes the correspondence between two texts, while parody emphasizes the difference. Parody's target is always another text or genre of texts (and so it tends to mock the form of these texts), while satire's target is something besides texts in the outside world (36–44).

11. Interestingly, many people who appreciate the film seem to have only the vaguest notion of the original Arthurian stories. In fact, one friend told me that most of what he knows about Arthur he learned from *Grail*. This reverses a basic assumption about the nature of parody: that one has to know the text(s) being parodied before one can make sense out of the parody. Since audiences who encounter the parody before they discover the original material can not only make sense but gain pleasure from *Grail*, this calls for a revision. Audiences can and do read parodies without knowing the original, although literary scholars may assume that one needs full knowledge to appreciate parodies.

12. For instance, see Landow (1992, 137–38) and Landow and Delany (1991, 6).

WORKS CITED

de Certeau, Michel. 1984. *The Practice of Everyday Life*. Trans. Steven Rendall. Berkeley: University of California Press.

English, David. 1994. Hot New Multimedia Products from Comdex. *Compute* 16, no. 9 (September): 61.

Gunning, Tom. 1986. The Cinema of Attraction(s): Early Film, Its Spectator and the Avant Garde. *Wide Angle* 8, nos. 3–4: 63–70.

Hewison, Robert. 1981. *Monty Python: The Case Against*. New York: Grove Press.

Hutcheon, Linda. 1985. *A Theory of Parody: The Teachings of Twentieth-Century Art Forms*. New York: Methuen.

Jenkins, Henry. 1992. *Textual Poachers: Television Fans and Participatory Culture*. New York: Routledge.

Johnson, Kim "Howard." 1989. *The First Two Hundred Years of Monty Python*. New York: St. Martin's.

Landow, George. 1992. *Hypertext: The Convergence of Contemporary Critical Theory and Technology*. Baltimore: Johns Hopkins University Press.

Landow, George P., and Paul Delany. 1991. Hypertext, Hypermedia, and Literary Studies: The State of the Art. In *Hypermedia and Literary Studies*, ed. Paul Delany and George P. Landow. Cambridge: MIT Press.

Rifkin, Jeremy. 1987. *Time Wars: The Primary Conflict in Human History*. New York: Simon and Schuster.

Smith, Evan. 1994. Python Bytes. *Texas Monthly* 22, no. 11 (November): 88.

Smith, Greg M. Forthcoming. Navigating Myst-y Landscapes: Utopian Discourses and Hybrid Media. In *Hop on Pop: The Pleasures and Politics of Popular Culture*, ed. Henry Jenkins, Jane Shattuc, and Tara McPherson. Durham: Duke University Press.

Chapter Four

"Evil Will Walk Once More"
Phantasmagoria—*The Stalker Film as Interactive Movie?*

Angela Ndalianis

"The Intertextual Arena" and Cross-Media Convergences

Two distinct tales of horror. Two heroines. Two psycho-killers. Two small-town communities. In the first story, the horror begins when a deranged murderer (possibly also the bogeyman himself) interrupts the peace of a small town. Lurking in the shadows, he emerges only to butcher a stream of unsuspecting young victims. At the end of the tale, the story's victimized and only surviving character, Laurie, rises to status of hero as she confronts the "bogeyman" head-on. Trapped in a house with him, her life balancing on a fine line, she has no option but to bring him out in the open and lure him to his own destruction. In the second story, the horror emerges when the heroine-to-be's husband develops psychotic, serial killer tendencies. The peace of their idyllic home and community is shattered and the psycho-killer's victim list builds up. Then Adrienne, the killer's wife, is left with no other option: she must engage him in final battle and, likewise, set him up for his own bloody annihilation. Two defeated psycho-killers. Two female victors.

Laurie and Adrienne's dilemmas and conquests sound like classic plot actions belonging to the stalker film tradition. A psychotic killer stalks members of a small community; a (usually) female hero is left as sole survivor of the story's main cast; and after a bloody and gory fest of mayhem and carnage, he hunts her down in an enclosed space (often a house), forcing her to meet his attack head-on. There is, however, one small difference: where Laurie belongs to the realm of the cinema, being the main protagonist in the stalker film *Halloween* of 1978, Adrienne belongs to the realm of the CD-ROM "interactive movie" and is the hero of the horror game *Phantasmagoria* of 1995. The exchange of character types, settings, sound effects, plot structures, and thematics present in these two media examples reveals the complex interchange occurring between contemporary entertainment industries.

At first glance, this exchange seems to involve the simple transfer of codes and conventions from the cinema into the world of the CD-ROM, but this "simple" migration entails complex transformations that unsettle film form and the interpretative approaches that are applied to horror cinema—and to film spectatorship in general. The first part of this essay will explore the intertextual and cross-media nature of contemporary entertainment media through an exploration of *Phantasmagoria*'s influences. In particular, I will draw on Carol Clover's and Vera Dika's examinations of the slasher/stalker film to emphasize the applicability of film models for CD-ROM interactive movies. Having acknowledged the convergence of two separate media by drawing on similar theoretical and critical models, I suggest that, despite the affinities shared by one genre across two media formats, CD-ROMs call for more fluid responses that can encompass the fluid and multilinear nature of this new technology. I will employ the concept of the hypertext as a means of examining the implications of multilinearity in *Phantasmagoria*.

As is the case with many examples of contemporary cinema, games like *Phantasmagoria* (much like slasher films) are aware of their own historicity. The intertextual referencing reveals a medium steeped in the history of its own conventions. When tackling *Phantasmagoria*, we need to consider how this CD-ROM game positions itself in the context of the array of conventions and traditions the game emerges from and consciously draws upon. As Jim Collins states in relation to the current proliferation of Batman texts, one characteristic of popular culture is its increasing "hyperconsciousness about both the history of popular culture and the shifting status of popular culture in the current context" (1991, 165). *Phantasmagoria* is no exception. An awareness of generic and stylistic antecedents is nothing new. The difference is that post–1970s examples place themselves in what Collins calls an intricate "intertextual arena." Focusing on the array of the "already said" (a term he extends from Eco), contemporary examples of popular culture not only draw on examples specific to their own genre in all stages of their historical development, they also reconfigure and negotiate the "already said" of popular culture, moving across different genres and different media, and revealing the fluid nature of the array (167).

While media examples like *Phantasmagoria* are the result of "multinational, profit-oriented media giants," we need to consider both the producers' and the audience's complex "intertextual frame," which is formed by a series of diverse sources (Pearson and Uricchio 1991, 2). Like the "bat-texts" that Collins speaks of, *Phantasmagoria*'s meaning depends on its negotiation of a variety of traditions past and present, including literary, cinematic, and computer games. From a literary perspective, in addition to the impact of a gothic sensibility, there are distinct echoes of the works of Edgar Allan Poe, in particular the haunting and eerie themes and atmosphere of tales like "The Fall of the House of Usher." Similarly, the small-town, familial crises that prevail in many of Stephen King's novels also find their way into the CD-ROM.

A precinematic heritage, evident in the game title's reference to the phantasmagoria, is also present. Phantasmagoria were especially popular in the late eighteenth and nineteenth centuries, and were included in the magician's repertoire. This precinematic entertainment form also revealed a predilection for themes of horror. The device, which relied on magic lanterns, could conjure illusions and phantasms that lured the audience into its horrific and fantastic representations.[1] In addition, the success of the phantasmagoria was further enhanced by the nineteenth-century love of things gothic (Heard 1996, 27), and its special effects and themes influenced early photography and the cinema—both of which dabbled in effects trickery through fantastic and horror themes.[2] The CD-ROM calls on this tradition not only through the game title, but through Carnovasch, the demonically possessed first owner of the house in *Phantasmagoria,* who was also a nineteenth-century magician versed in the "magic" of the phantasmagoria. As such, the CD-ROM game *Phantasmagoria* is acknowledging its history, acknowledging the precinematic impact on horror cinema, and acknowledging its own place in redirecting this horror tradition into new, computer game directions.

Traveling through diverse media such as print media, the precinema, and the cinema, the game also produces meaning by interacting with its computer game heritage. In the case of computer games, however, rather than being influenced by horror games alone, *Phantasmagoria* turns to other CD-ROM games like *Under a Killing Moon, Gabriel Knight, Return to Zork, Critical Path,* and *The Seventh Guest* for the interface, the use of full video motion, the ability to move the characters, and "filmic" presentation of events; and to games such as *Myst* for the depiction of dazzling and haunting computer-generated environments. The result of this complex web of intertextual references is that the narration is not simply limited to completing the plot, or to becoming involved in the "syntagmatic axis of the narrative" (Collins 1991, 168). Instead, the "layering of intertexts that occurs simultaneously deforms those same topoi along a paradigmatic axis of antecedent representations" (Collins 1991, 169). The "hyperconsciousness" permits us to become engrossed in the narration in a more conventional sense, with the plot and themes unraveling along "syntagmatic" lines, but the viewer/player is also encouraged to extract meaning that lies in the audience's awareness of the multilayered, intertextual references (Collins 1991, 173). Story concerns, therefore, are not the primary drive; allusions to other media examples that have had an impact on *Phantasmagoria*'s construction of meaning are just as integral to the player's involvement in the game. We see this fluid, intertextual motion when we consider *Phantasmagoria*'s technological, hypertext status. Before dealing with the concept and technology of the hypertext, however, I will now turn to the horror intertext.

Stalker Film Meets Stalker CD-ROM "Interactive Movie"?

Recently horror has found a popular and lucrative home in the CD-ROM format. Horror film conventions have migrated and mutated into computer games like *Phantasmagoria, Phantasmagoria 2: A Puzzle of Flesh, Harvester, Ripper,* and *Alone in the Dark I, II* and *III. Phantasmagoria*—which was created, written, and designed by Roberta Williams and directed by Peter Maris—has proved to be one of the most controversial games released. The game's horror themes and gory displays of violence, while common to the world of horror cinema, were unprecedented in computer games. In terms of the new representational realism provided by the technology that is the CD-

ROM interactive movie, *Phantasmagoria* was groundbreaking. In fact, it was the combination of realistic imagery and interactive technology in the context of horror that proved a controversial issue among critics who viewed the game's gruesome and realistic effects as capable of producing real-life incidents of horror.

Censorship and controversy aside, the game made advances in the interactive potential of horror games by abandoning the animated environments (familiar to games like *Doom, Alone in the Dark,* and *Quake*), incorporating instead real actors (twenty of them), settings, and cinematography and stylistic devices traditionally associated with the cinema, including five hundred camera angles. In the game, not only is it possible for characters to move around a realistic 3-D environment—aided further by the incorporation of full-motion video—but the level of character motivation and character development had not been witnessed before in computer games. The recent surge of horror games like *Phantasmagoria* actually allows the game designers to showcase the new, creative and atmospheric potential of computer games; this is made possible by CD-ROM technology's capacity for producing advanced, high-resolution graphics (Foster 1995, 32). The production of a game like *Phantasmagoria* also reflects a recent trend that involves the blurring of boundaries between making a film and producing a computer game. As a result of this shared generic, stylistic, and intertextual experience, to a certain extent it is useful to consider the horror games from the perspective of horror cinema. However, as will be discussed later, more flexible forms of interpretative models need to be incorporated—models that take into account both film traditions and the different interactive possibilities provided by CD-ROMs.

Parts of the game—especially chapter introductions (of which there are seven), flashback fantasies, and nightmares—are articulated in a movie format (in full-motion video); in this instance it isn't possible to manipulate the action unraveling before us. The rest of the game, however, is interactive. By using our mouse we direct Adrienne to make explorative choices that slowly untangle the narrative web mapped out by the game designer. These two modes of representation come together to produce an interactive environment that allows the player (who is responsible for guiding Adrienne through the game) to move around in the house, its gardens, and the town by clicking in the direction we want Adrienne to take. It is possible to examine and collect various objects along the way (which are saved in an inventory), and which are put to use at later stages in the game. For example, a letter opener is retrieved from the inventory to remove a loose brick in the fireplace; this allows Adrienne access to a mysterious chapel that has suspiciously been sealed off from the rest of the house. A bone that Adrienne collects from the general store permits her to coax the ferocious dog guarding Malcolm's house, and therefore gain crucial information about the history of the horrors that have befallen Adrienne and Don. And a Christmas tree snowman plays a fundamental role in the demise of mad Don at the end of the game.

Phantasmagoria shares many generic concerns with evil possession films like *The Shining* (1980) and *Amityville Horror* (1979). The couple, Adrienne Delaney and Don Gordon, arrive at their new gothic-style home in the small town of Nipawomsett. While exploring her new home, Adrienne discovers a secret chapel hidden behind the fireplace in the library. In the chapel she unwittingly unleashes an evil demon, who possesses her husband, Don; his frame of mind slowly shifts from testy to unpredictable to deranged. In attempting to unravel the secrets of the house, Adrienne discovers that a century earlier the house was inhabited by the magician Zoltan Carnovasch—otherwise known as Carno. Finding the trickeries of magic insufficient to quench his magician's thirst, Carno turned to the black arts in the hope of producing "real" magic. Purchasing a spellbook that opened the way to the gates of hell, he summoned an evil being into our world. In unleashing the demon, Carno's spirit became possessed, with the result that over the span of a decade he married then brutally murdered his five wives. The last one, Marie, plotted with her lover Gaston to murder Carno by trapping him in one of his magician's traps. The result, however, was that all three died, finally putting an end to the evil in Carno's host body—that is, until the arrival of Don. Mimicking his soul mate of the nineteenth century, Don proceeds on a bloody killing spree that puts an end to most of the supporting cast—including Adrienne's cat, Spazz. Thrust unwillingly into a state of horror and disbelief, Adrienne discovers—one by one—the dead bodies of Don's victims. She also witnesses the nineteenth-century phantom slayings of Carno's five wives, who he disposed of in inventively gruesome ways.

While the game owes a great deal to possession films, it is the game's links with the stalker tradition as portrayed in films like *Hal-*

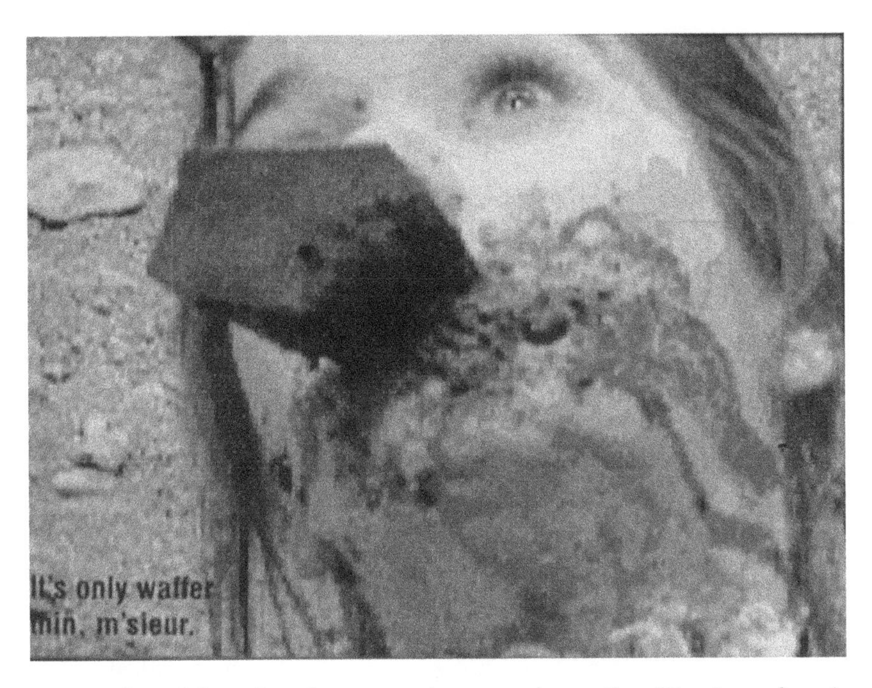

Fig. 4.1. One of Carno's unfortunate wives experiences the ultimate gardener's nightmare: a garden trowel thrust viciously into her mouth by her possessed husband.

loween, Prom Night (1979), and *Friday the 13th* (1980) that remain my primary concern. Both Vera Dika and Carol Clover have outlined the basic formula of these popular examples of the horror film. In *Games of Terror*, Dika provides an analysis of the stalker film, which she suggests dominated in popularity between 1978 and 1981; the most influential instigators of the formula were John Carpenter's *Halloween* and Sean Cunningham's *Friday the 13th*. Her formula focuses specifically on a combination of plot, setting, character types, stylistic components, and production values that dominated and were obsessively repeated in the subgenre during this period. Clover, in *Men, Women and Chainsaws*, dubs these films "slasher" films and extends the temporal limits to include earlier examples such as *Texas Chainsaw Massacre* (1974) and *Texas Chainsaw Massacre 2* (1986). Rather than focusing on Dika's more formalist concerns, Clover's slasher definition emphasizes the dynamics centered around the "slash" and the relationship that eventuates between psycho-killer and what she calls the "final girl," who is the final survivor and hero of the films. Despite differences of opinion between the two writers, the fundamental logic of the stalker/slasher film remains the same: it concerns shifts of power and, in particular, the narrative (and spectatorial) reasoning behind the shifts of power away from psycho-killer to final girl hero. Four elements present in this subgenre of horror are also applicable to *Phantasmagoria*: plot structure, character types, structures of the gaze, and the gore factor.

Dika locates a dominant pattern in the stalker film. The narrative divides into two separate temporal frameworks: one set in the past, and the other in the present. Yet both are linked in the way the killer (whose initial crime is depicted in the past) and the victim/hero (who we are introduced to in the present) are to be brought together thematically in the film's present temporal frame (1990, 59–60). *Phantasmagoria* does operate on this past/present principle. However, rather than demarcating abrupt structural separations between past (which frames the opening) and present (which constitutes the bulk of the narrative), the game intersperses the past throughout the game. The present scenario of Don and Adrienne is intermingled with the past, which consists of a series of flashbacks that appear to Adrienne as phantom visions of Carno's wives appearing to "retell/show" their gruesome deaths. Despite the tampering with the sequence of events—which, as will be outlined later, is the result of the game's technological nature—past crime still provides an answer of sorts to the heinous crimes taking place in the present.

Phantasmagoria's other point of contact with the stalker film is its attitude to gory special effects. The stalker/slasher tradition parallels the birth of gore in contemporary horror cinema. It was the stalker/killer's murderous acts that sparked horror's infatuation with splatter and a "buckets of blood" attitude to special effects. In the many moments of bloody destruction, gore-effects often compete with narrative progression and character motivation. *Phantasmagoria,* as inheritor of this tradition, is no exception. Hortencia, Carno's first wife, dies in her greenhouse: we look on with Adrienne as Carno thrusts a garden trowel into her mouth, slitting her lips further apart into a Joker grin, and continue to watch as he stuffs her morbidly wide-grinning mouth with soil. Victoria, his second wife, had a predilection for the vino bianco—and Adrienne/we look on aghast, intent on every gory detail, as Carno ruthlessly plunges a wine bottle into her eye socket. And at the hands of a bad game player, we even see Adrienne herself sliced right through—head and torso—by a giant axe, her blood and flesh splattering outwards and filling the inner space of our computer screens. Such scenes of vivid, horrific violence would find a welcome niche in splatter classics like *Friday the 13th*, *Happy Birthday to Me* (1981), and *Slumber Party Massacre* (1982). During those minutes of splatter-fx, narrative concerns are put to one side, and the focus becomes the display of special effects themselves. Interestingly, due to censorship issues and the controversy that surrounded the game, the designers included a button that permits the player to "censor" the game (and therefore remove the display of the violence and gore-fx scenes). Of course, the censor button was introduced to appease the censors—the assumption being that the conscientious and morally bound game player (or parent of a game player) would opt to censor the game of its grosser and more vivid moments. In some respects, however, to put this censor button into action (and which game player in their right mind would?) is to deny the game its place in this post-1960s splatter horror tradition.[3]

With regard to the spectator's investment in the slasher film, Clover argues that these films are more victim-identified than traditional film theory has allowed for. Rather than operating according to a voyeuristic gaze that is about control and mastery, the spectator of the slasher film responds to and "identifies" with the main character—the final girl.[4] In the process he (and it is the male spectator who interests Clover) becomes embroiled in a position that, for most of the film, thrusts him into a state of powerlessness in the face of evil (Clover 1992, 8–9). It is only in the final stage of the narrative that the powerlessness shifts to a position of control through the female protagonist's transformation into hero and active instigator of narrative action.

Clover explores Jurij Lotman's analysis of character functions in myth and arrives at a similar conclusion: the character functions of the victim and hero are gendered (1992, 12). Like myth, slasher films reveal only two character functions: society codes mobile, heroic functions as masculine, and immobile, victim functions as feminine (13).[5] Arguing that the slasher film reflects an older tradition that saw the sexes not in terms of differences but in terms of sameness, she suggests that the main characters (hero and killer) encapsulate the one-sex model. Both sexes (and their accompanying functions) are contained in the one body. *Phantasmagoria* fits quite neatly into this argument.

Secondary characters, for example, exist in the slasher film in order to be "killed off," their status as "function" remaining in the victim/feminized mode.

These "feminized"-victim characters—who included the vagrants Harriet and Cyrus, the telephone repairman, even the five murdered wives of Carno the magician—serve a narrative purpose in highlighting Adrienne's heroic properties when they finally emerge. Unlike the characters doomed to be victims, in the final girl and psycho-killer both sexes (and their accompanying functions) are displayed. As the killer moves toward defeat he becomes feminized, and as the final girl moves toward victory she shifts from role of victim to that of masculine/hero (Clover 1992, 50). Dika concludes in a similar fashion in stating that both the killer and the heroine are masculine in their "dominance of the film's visual and narrative context" (1990, 55).[6] In all, the complex processes of character empathy or identification engaged with in the slasher film reflect the fact that identification in the cinema is a fluid process, one that need not rely solely on same-sex identification, or on an equation that correlates the male gaze of the spectator to one always centered around mastery and control.[7] As Clover states, the slasher film reveals a fluidity and theatricality of gender often ignored by traditional film spectatorship theories.[8]

The theatricality of gender is highlighted further in *Phantasmagoria*. Adrienne's victim status in the initial stages of the game is quite typical of the final girl. Her suffering at the hands of her victimizer is seen, for example, in the final chase scene through the house, and in the noninteractive rape scene that takes place in the bathroom in chapter 4. The absence of interactivity in the presentation of this scene further highlights her passive victim status—a passivity also experienced firsthand by the player in his/her inability to interact with the dramatic action. But Adrienne's passive elements are always doubled (and finally overshadowed) by an assertiveness and refusal to slip unquestioningly into the role of victim. Despite Don's threatening ravings and growing violence, Adrienne remains inquisitive, assertive, and active—traits that allow her to take on the status of hero in the end. Likewise, in the final "Don confrontation scene," the full-motion video mode is abandoned; Adrienne's greater active, heroic function is asserted by our capacity to now interact with the events. In addition to the "one-sex" character function that sees Adrienne transform from victim/feminized function to hero/masculinized function (with Don following a reverse transformation), similar fluidity is witnessed in the possibility of cross-gendered identification that is provided to male game players. As with the slasher film, since the final girl is the only fully developed character, and the one who drives the game narrative forward, the male player has no option but to "identify" with Adrienne's character. In fact, the investment in the female character is more prominent in the game than in film examples. A film unravels itself in a linear manner beyond our control, and the viewer can choose to reject identification with the main protagonist. The viewer, for example, may choose to "identify" with the ritualist and highly repetitive narrative conventions of the stalker film, or with the subgenre's equally ritualized displays of gory special effects—both of which take attention away from primary focus on the main character. However, the point-of-view structure of the games makes identification (in the sense of narrative interest) with a character more crucial. On a fundamental level, the game would not progress if the player chose not to "play the part" of Adrienne.

The final girl is the "undisputed 'I' on which horror trades" (Clover 1992, 45), and her central role is reflected in the camera work that the films use to construct point of view shots. The construction of point-of-view shots is more crucial to the interactive environment of CD-ROM technology. The stalker/slasher tradition favors combining more objective, third-person shots with the subjective points of view of both killer and hero, where the camera often represents the vision of these characters. As Clover points out, the "eye" of the camera is collapsed into the "I/eye" of these characters. While the killer's point of view dominates in the first part of the film, in the latter part (when the final girl shifts from victim mode to that of hero), she gains control of the gaze. *Phantasmagoria* reflects a similar structure of gazes that unravel and transfer the power of the gaze as the narrative progresses. However, as will be discussed later, the strict linearity of this unraveling need not necessarily be adhered to. The game presents the player with four structures of the look. The first two are more interactive modes involving first- and third-person views—usually in combined form. For example, Adrienne's examination of environments (seen in third-person, more objective mode) often also leads to a more subjective shift where her vision merges with that of the camera/computer eye. For example, we look on as Adrienne moves toward the window

Fig. 4.2. An objective third-person view of Adrienne as "final girl" attempting to solve the mystery.

in the tower room, then our viewpoint is collapsed into Adrienne's/the camera's as she/we look through the window toward the greenhouse.

The two final structures are characterized by a "movie"-style (and therefore noninteractive) cinematography that is again both objective and subjective in method. More objective, third-person camera work includes the opening scenes to each introduction. The more subjective shots, on the other hand, reveal the layered complexity of this intertextual array, including the so-called fly-by shots, which are rapid tracking shots (where the camera reflects the player's/game character's vision) made popular in the computer game world by the mystery/horror game *The Seventh Guest*. Interestingly, the "fly-by" shot finds its most famous film parallel in the stalker film, particularly in the articulation of the killer's point-of-view shot as he (and sometimes she) stalks his victims. *Phantasmagoria* acknowledges this tradition. This cinematographic convention as staple of the subgenre was initiated by *Halloween*, finding almost poetic expression in Michael Myer's stalking, invasive gaze. However, later in the film, just prior to Laurie's discovery of the blood-fest (and as she crosses the street to the house of death), she is also associated with this viewpoint—and its accompanying associations with control of the gaze and narrative action. *Phantasmagoria* also permits us access to Don's and Adrienne's visions through this more subjective camera work, although, interestingly, the game tends to favor Adrienne's gaze. True to the stalker/slasher subgenre, Don's gaze is privileged when the game emphasizes Adrienne's impending victim status. For example, in chapter 3, after Adrienne returns home from town (and after Don has undergone the early stages of his nasty transformation), she arrives in the BMW, parks the car, gets out, and a "fly-by" tracking shot begins, quickly and menacingly making its way toward her. Adrienne's viewpoint then takes over, and we discover that the menacing gaze, which (true to the *Halloween* tradition) had been collapsed into our own, had been Don's.

The game, therefore, follows the structures of the look according to the slasher/stalker tradition, and ultimately it is Adrienne's gaze that remains prominent. Adrienne's point of view is framed by two kinds of shots. The first is the tracking shot that reflects the operations of her mind's eye, the most powerful instance being the opening nightmare "scene" that has the first-person eye/camera eye flying through a dark, atmospheric environment filled with torture instruments and horrors of all kind. The tracking viewpoint travels restlessly past surreal, abominable images, then stops suddenly as viewpoints change and Adrienne looks on at her own torture—a foreboding warning of the chaos that lies ahead. The second is a point of view more traditionally associated with the stalker film; for example, the tracking shot that follows her gaze as she glances down the second floor staircase. However, the speed of the tracking point-of-view shot implies that it could in no way be associated with a character's vision. As mentioned earlier, the complex intertextual play entails the game situating itself in the context of the horror tradition "array." The result is that the structures of these looks may imply a direct point-of-view shot associated with the late 1970s stalker film, but the shot also extends itself, replacing character point of view with camera point of view. This subjective shot depicting camera in frenzied motion is a convention developed by Sam Raimi in his *Evil Dead* (1983), a film that took to excessive, parodic limits the conventions of the stalker film. The game, therefore, creates links with aspects of the stalker film across its history (as well as horror conventions of computer games like *The Seventh Guest*), and this further complicates the saturated quality of allusions and intertextual references present in the game.

The Hypertextual Array

Such interpretations that draw on film criticism and theory are important in placing CD-ROMs like *Phantasmagoria* in the context of their intertextual filmic and generic traditions. However, in exploring the web of influences, we must also consider the significance of the generic migration of horror into the realm of the computer. Clearly Clover and Dika never intended their interpretations to be applied to CD-ROM games, but the issues they raise are significant, given the interrelationship of two media forms bound by similar generic concerns. However, when drawing on film critical models, critics will need to transform their interpretations in order to accommodate the radically different narrative forms suggested by CD-ROM game technology.

The convincing application of Clover's and Dika's approaches to *Phantasmagoria* would depend on the fixity of the text—a text that relies on a story progression that shifts in a linear fashion from problem to resolution. In both instances, the static, linear structure of the stalker/slasher film allows these writers to view the films' plot structures (and the stylistic codes that support the plot structure) as a membrane. When delving below the surface, one exposes the "meaning" of the text and along with it its social, psychological, or mythic significance.[9] In Clover's case, for example, much is to be gained by a male spectator identifying with a female hero. Since the film "run[s] identification through the body of a woman" (who serves a dual function as victim-hero), castration anxieties associated with the loss of power, and with "being feminized" find a safer outlet in the female hero. The final girl functions according to a "politics of displacement" and the woman is used as a "feint, a front through which the boy can simultaneously experience forbidden desires and disavow them on the grounds that the visible actor is a girl" (Clover 1992, 18). For Dika the repetitive, formulaic features of the stalker film—and their peak popularity between 1978 and 1981—reveal a ritualistic logic. The films engage the audience in a game about genre and the conventions of the stalker film, the underlying intention being a social ritual that addresses a collective dream. On a psychoanalytic level, the Oedipal conflict underlying the film narratives addresses the needs of adolescent spectators (Dika 1990, 16). On a broader social level, the films provide an outlet from social traumas such as post-Vietnam anxieties, inflation, unemployment, and the like (Dika 1990, 131).

Indeed, the CD-ROM "interactive movie" has been criticized for its greater emphasis on linear form, especially when compared to games like *Quake*, *Alone in the Dark*, or *Duke Nukem 3D*. Unlike *Phantasmagoria*, the latter are not dependent on video imaging and a game design that requires the use of real actors, film cinematography, and a presentation more in common with television and the cinema. The CD-ROM "interactive movie" format, which is still in its early stages, is seen as limiting the player's interactive and participatory options. Gary Penn, for example, has complained that the detailed environments of CD-ROM interactive movies limit the interactive potential of the game, making the games more closed and linear; for Penn, "linear footage is not the way forward in creating detailed interactive environments" (1994, 86). However, as Keith Ferrell states,

> On a purely pragmatic level, the creation of open-ended, wholly interactive, fully explorable worlds that still possess some sort of structured story and character content may be too much to ask. . . . How do you anticipate every possible scenario or player action? How do you ensure complete narrative consistency no matter what the actions are? How many millions of words of dialogue must you write in order to accommodate all the conceivable conversations? Add the creative, technological, and budgeting challenges and you're looking at an undertaking that dwarfs even the biggest of motion pictures. (1994, 30)

A more participatory and interactive mode of presentation encompasses incredible creative and technological demands. Yet despite the logistical limitations, a strict linear structure is nevertheless ruptured, and we are presented with possibilities totally alien to a cinematic experience. Like many CD-ROM games, *Phantasmagoria* unsettles traditional modes of interpretation, operating instead according to the logic of interaction—specifically interaction that depends on hypertext, web-like narrative forms that present multiple rather than singular narrative possibilities. While the seven chapters of the game remain more or less linear, in each chapter multiple plot directions are presented to the player depending on the choices he/she makes. Therefore, despite clear parallels between cinema and computer game genres and themes, we need to make adjustments when applying film models to computer games.

Espen Aarseth has noted that the development of the computer as both cultural and aesthetic form of expression challenges the paradigms of cultural theory (1994, 51). When film experience is applied to computer game experience, the linear, closed, and passive should now be replaced by the multilinear, open, and interactive. Interactivity is made possible by hypertext technology that calls into play alternative metaphors to those traditionally conceived for the film spectator. A hypertext system consists of a series of blocks or "lexias" that are connected "by a network of links or paths" (Snyder 1996, 49). In the case of *Phantasmagoria* each block presents only one possibility of a narrative sequence; a variety of story possibilities branch off into numerous directions. The narrative is not limited to a static, spatial fixity; instead, it is an open, dynamic structure that moves beyond the linear. According to Ilna Snyder, just as the technology itself abandons fixity through its hypertext links, so critical theory must follow suit, and the new metaphors must "evoke features of hypertext-technology." The three most familiar metaphors for the new experience of narrativity are navigation, the labyrinth, and the web—and *Phantasmagoria* realizes all three metaphors.

We, for example, are responsible for making decisions as to how and where to guide Adrienne. As navigators of the story, we select the sequence of events and the sequence in which the narrative unravels before us. Certainly, reflecting the structure of the labyrinth, some aspects of the narrative action are closed off to us until we have explored others; this labyrinthine format also implies—to a certain extent—a linear attitude to narrative form. For example, it is not until chapter 5—after the seance held by Harriet in which Carno appears to Adrienne—that we can operate the dragon lantern in the conservatory; this, in turn, allows us to access a secret passageway in the room, which leads us to the previously out-of-bounds theater. However, we are still instigators in organizing narrative events in a way that is unique every time we restart the game. The overall effect is the creation of a web of plot scenarios that undermine notions of singular, linear progression. For example, while in the kitchen, we/Adrienne may trek off to explore Carno's bedroom (where Adrienne later lies on the bed and has a nightmare sequence about multiple, decaying hands emerging from the bed to capture her). But to get to the bedroom we may choose to go via the dining room; or first go outside and explore the gardens; or go through to the adjacent bedroom that belonged to Carno's last wife, Marie, and so on. Each decision, of course, is accompanied by its own narrative implications—and conceivable interpretations.

If we choose to apply both Dika's and Clover's arguments to *Phantasmagoria* we need to reevaluate the conclusions they draw based on film texts. In the crucial shift in the relocation of power away from the psychotic killer and to the final girl, the shift from victimization and passivity to mastery and control can alter dramatically. Certainly, if we play the game "properly", ensuring Adrienne's success in the end, we could argue that Dika's and Clover's assertions are still applicable. Regardless of the hypertextual structure, in one sense there is still a core, binding narrative that we are required to play. And chapter 7, the final chapter of the game, may unravel to reveal classic "final girl" material. Don corners Adrienne in the dark room, she (and we, her alter ego) struggles with him, then she ingeniously chooses to throw drain cleaner on his face. Escaping to the nursery, she finds a shard of glass from a broken frame, and hides behind the door waiting for Don to come. When he does, Adrienne plunges the broken glass into him, then makes a run for it into the theater dressing room downstairs. She finds Don's jacket on the floor, searches it, and discovers a Christmas tree decoration in the pocket (which has sentimental value because Don proposed to her on Christmas Eve). Hiding in the closet, she waits out her fate. Don enters soon afterwards, opens the door, and drags Adrienne to the chair of torture. He manacles Adrienne's right hand to the chair and, as he leans over to manacle her feet, she/we select the snowman from the inventory and give it to Don, as she pleads with him to remember how things once were between them. As Don begins to lift himself from a crouching position, with her free left hand we make Adrienne pull a lever that is situated to her left. This lever sets the wheels in motion: the giant blade situated above them—which was meant for Adrienne—plunges explosively and memorably through Don's upper body.

According to one of the hypertext series of links, this narrative possibility—and the interpretations that accompany it—is still viable. However, it remains only one of the narrative possibilities present in the game. This narrative ending depends on us playing the game through without making mistakes, remembering to search through environments properly, discovering all necessary bits of information, and picking up all useful objects. The above mentioned linear scenario is but one of many web-like plot formations that can envelop us. The unraveling of the game depends on "traversal functions" (Aarseth 1994, 61) that exist across blocks within the hypertext link. According to hypertext theorists such as Aarseth, the concept of the hypertext entails nonlinear structures. However, despite this seeming randomness and lack of linearity, when we play a game it unravels before us in time; as Liestol argues, the possibility of the nonlinear in time is an imaginary construct that is impossible because "time is linear" (1994, 106). Liestol continues, "However discontinuous or jumpy the writing or reading of a hypertext might be, at one level it always turns out to be linear . . . nonlinearity exists only as positions *in space*, different alternatives of which one may choose only *one at a time*. . . . Nonlinearity, one might say, is never actually experienced directly" (106–7). What Liestol suggests, therefore, is the formation and experience of multiple rather than singular narrative strands that branch outwards. And with each new strand of the narrative web we, in a sense, need to alter our interpretation to follow suit. When I played *Phantasmagoria* a second time, I searched the theater dressing room in a different sequence—before Don had visited and left his jacket. The result was that Adrienne was strapped to the blade-chair without the snowman. No snowman, no way of coaxing Don to strategically position himself in the line of the giant axe. No snowman, and the final girl becomes final in the sense that she dies: the axe plunges mercilessly through Adrienne's body, narratively leaving her in the position of feminized/victim. In such a scenario, when applied to the CD-ROM game, Dika's and Clover's assertions about mastery at the end no longer remain fixed, and conclusions need to be reevaluated. Our critical models and methods of interpretation need to be just as hypertextual and dynamic as the narrative webs the games present.

Phantasmagoria can further unsettle the static quality of the linear text, taking multilinearity in yet more directions. For example, we can access video files (*.VMD, "Video and Music Data" files) from each of the seven CD-ROMs, and copy them onto our hard drive. This allows us to view these scenes at different temporal points in the story. For example, Don's rape of Adrienne in chapter 4 can be shifted into an earlier chapter. This can radically affect questions of character motivation—especially if the rape takes place prior to Don's demonic

Fig. 4.3. Adrienne as literal "final girl," her head sliced by an axe—compliments of an inept game player.

possession (implying different kinds of narrative motivations that would suggest that the evil events that took place were reflections of a problematic relationship that existed between Don and Adrienne prior to the release of the demon). Likewise, those not interested in the narrative elements at all can copy all the *.vmd files onto their hard disk. This option is a gore and effects junkie's dream because the fx scenes of gore and grueling violence can be accessed at will—and removed from any narrative logic. In this instance, the game "meaning"—and the linear path we choose to create—focuses around the unraveling of special effects spectacle. As a result, our critical response to it must follow suit in that, in this instance, narrativity is put to one side and the spectacle of special effects dominates as source of meaning.

Horror cinema is renowned for its obsession with rupturing the closed, linear structures of its film spaces—particularly through its endings that refuse the closure of many Hollywood films—and *Phantasmagoria* is clearly aware of this convention. For example, after a possible stalker/slasher ending (and assuming Adrienne's successful destruction of Don), Adrienne moves on to a second ending, one that addresses itself to the possession/evil spirit tradition of horror. When she kills Don, the spirit that possesses him drifts from his body in a green mist, then solidifies into an enormous blue monster who hunts her down through the house. Again, Adrienne's success depends on how we play the game. If we run to the chapel and use the spellbook (which we should have picked up in the dark room), then chant an incantation after sacrificing some of our blood by pricking our finger with a broach we should have in our inventory, the demon will be trapped, unable to harm anyone again. But if we fail to do any of the above, again we shift to loser status and the game ends with Adrienne dead and us anything but victorious.

This double climax is a convention of horror cinema: one ending appears to end, while another opens up. In *Halloween*, for example, John Carpenter made popular the classic ploy of opening up a seemingly closed narrative structure. The story of the demise of Michael Myers appears to end (and he appears to be dead), then the final sequence of the film opens up this resolution by presenting the audience with Michael's disappearance. The cinema can certainly manipulate and rupture closed narrative form, suggesting the multiple narrative possibilities found in the hypertext. Similarly, the intertextual nature of genre has parallels with a hypertext structure in suggesting links that move beyond the limits of the enclosed text and across popular culture.[10] However, where *Phantasmagoria* is different from the cinema is that much of its intertextual and hypertext structure is the result of the technology that drives it. The hypertextuality of CD-ROM technology realizes the process of intertextuality in more literal, technological terms. In the case of the horror double climax in *Phantasmagoria,* multiple options are open to us as a result of the fact of the medium itself: we may never get beyond the first ending; we may get to the second ending but fail to complete it successfully; or we can reign victorious and survive both endings. Whatever the scenario—and no matter how "open" the final ending of a film may be—in hypertext fiction no final, static version exists that can be replayed in the same way as film over and over again.

The multilinearity and open narrative form can lead us into all sorts of other directions. If linearity and character motivation traditionally associated with Hollywood cinema are not the player's predilection (or if the player simply can't work out how to access all options present in the game design), it is possible to produce a narrative that has more in common with art cinema and avant-garde cinema forms of narration and nonnarration. In the case of bad game play, not only is it possible to miss entire portions of narrative action (thereby creating narrative gaps), but it is also possible to focus on actions that are in no way concerned with unraveling a narrative. We can make Adrienne wander aimlessly around the house and in town; we can make her eat in the kitchen or look at herself in the numerous mirrors littered around the house; she can go to the bathroom, comb her hair, put on makeup, and go to the toilet. In the last scenario, the usual procedure is that we click on the toilet seat, she shuts the door on our face, and then we hear the flushing sounds of the toilet, and water running in the hand basin before she opens the door. But one of the cheat-tricks fans of the game have discovered is that if we click seven times on the chair next to the toilet before clicking on the toilet, Adrienne shuts the door and we hear all sorts of gross sound effects—moans, groans, plunks, and "aaahs"—coming from inside as Adrienne goes about her business. All these actions—from the mundane to the comical—are more aligned with the "dead time" of art cinema. They are actions that are not concerned with progressing narrative action. In many of these nonincidents I was reminded especially of the mundanity of everyday life represented in Sally Potter's feminist avant-garde film *Thriller*. Despite Liestol's perceptive comments about the multilinear as opposed to nonlinear in the hypertext, these moments of "dead time" do emphasize nonnarrative patterns. While events such as these do unravel in a sequence that mimics the linear, they remain nonnarrative in the sense that they serve absolutely no story function.

Landow has suggested that the hypertext redefines the beginnings and endings of linear narratives in nonlinear—or, rather, multilinear—ways (1994, 34–36). In the case of computer games, not only are we confronted with multiple story avenues in each game session, but every time we recommence a game we are, in a sense, starting another beginning. In addition, as the above examples reveal, the hypertextual embraces multiplicity in other ways, allowing *Phantasmagoria* to unravel along the lines of Hollywood narrative form (with emphasis placed on cause-and-effect structures and character motivations); art cinema narrative form (which loosens up linear, causal relations); or avant-garde cinema nonnarrative structures (in the focus on "nonincidents" that undercut narrative concerns).

Depending on the game player's choices and interests, therefore, the possibilities are as labyrinthine as the hypertext structure that supports them. Each game player can alter the order and articulation of events in distinct ways. Even if other games similar to *Phantasmagoria* emerged, following along similar slasher paths (for example, its sequel *Phantasmagoria 2: A Puzzle of Flesh*), the intertextual array, interactive possibilities, and hypertext form would shift and change so dramatically that no theory or critical approach predicated on homogeneity and stability could be applied in any sustained or beneficial manner. As Snyder suggests, the hypertext "discounts any final version" (1996, 57). The search for singular interpretations in a sea of hypertext links becomes nearly impossible; the original text is "no longer inviolate" (Snyder 1996, 76). Not only are traditional conceptions of narrative and the original unsettled, but assumptions of authorial control also stand on shaky ground. While the game programmers are the creative voice in that they design the game structure and devise all possible hypertext scenarios, the game player also has an integral creative role to play in navigating and selecting the narrative (or nonnarrative) paths that give life to the game action in configurations specific to his/her methods of game play. An "original" version of a game is rewritten each time it is played and replayed. In the words of Aarseth, "As long as we are able to imagine or reconstruct an ideal version, everything appears to be fine, and our metaphysics remains intact. But what if the flawed version interferes so deeply with our sense of reception that it, in more than a manner of speaking, steals the show?" (1994, 56). What if, indeed? What happens when traditional narrative structures that theories have depended on are removed? Is the auteur truly dead? Is the original forever lost? Is the ideological purpose of popular narratives removed? Is their mythic function nonexistent? Perhaps, as Foster suggests, these new technologies (and changing forms of popular culture) are merely rearticulations of past forms, which nevertheless still serve a social purpose. Previous "tales of horror and fear . . . dealt with our basic human concerns—survival and protection against what was an uncontrollable world. Now . . . although modern society buffers the human race against the horrors of nature, we still crave the experience of controlled fear, induced for entertainment" (Foster 1995, 33). Is *Phantasmagoria* an example of such entertainment? Perhaps. Whatever the case may be, the new critical and theoretical approaches that we refine will need to be as fluid and multifarious as the media examples we analyze.

NOTES

1. Philipstal's *Phantasmagoria* shown in Edinburgh in 1802 included the fantastic horror effects that revealed the sudden apparition of ghosts and skeletons (see McGrath 1996, 15). The most famous presentation of the phantasmagoria was the *Fantasmagorie* by Étienne Gaspard Robert (known as Robertson). In 1797 Robertson performed the *Fantasmagorie* in Paris in an abandoned chapel surrounded by tombs. Crowds flocked to the dimly lit tombs to see magic lantern effects that included skulls, atmospheric lighting, sound effects, and ghostly apparitions (see Barnow 1981, 19).

2. On early photography and cinema's relationship to ghostly imagery, see Gunning (1995).

3. It is conceivable that the censor button can also function as a CD-ROM game variation of the spectator who closes his/her eyes at the scary and bloody sequences in horror cinema. Of course, this is an option also open to the game player.

4. On the theorization of the spectator along the lines of gendered identification and male mastery, see Mayne (1993), especially chapters 1 and 2.

5. Clover admits that some slasher films do have male protagonist heroes; however, in these rare instances—*The Evil Dead* (1983) being one case—the male character retains elements of both masculine and feminine traits.

6. There are, of course, problems with this almost essentialized correlation of masculinity with the active and heroic. For the sake of argument, I am interpreting Dika's and Clover's models as functions imposed by social convention.

7. I am aware that the question of identification in the cinema is a complex and problematic process. Rather than using the term according to the psychoanalytic model of identification as posed by post-1960s film theorists like Christian Metz and Laura Mulvey, I am using the term to suggest that the films (and games) present their final girl as a character to identify or empathize with, in the sense that this character remains the primary narrative focus, and we are invited to invest our interest in her.

8. See especially the "Introduction: Carrie and the Boys" and chapter 1, "Her Body, Himself." Laura Mulvey would, of course, be the most influential proponent of more traditionally aligned spectatorship models, arguing for the hierarchization of gendered looks and power structures; the male spectator is placed in a position of spectatorial control that allows him to identify with the

dominant, diegetic male characters—and thus reaffirming the passive nature of the female character (see Mulvey 1975). Even her revision of same-sex identification reinstated this power structure. While the female spectator was acknowledged, Mulvey argued that she was often forced into cross-gendered forms of identification in that this was the only option that allowed her narrative control akin to that of the male spectator (see Mulvey 1981). The possibility of the male spectator opting for similar cross-gendered forms of identification was not considered. As Clover argues, the fascinating feature of the slasher film is that it invites the male spectator to identify with characters who fail to submit to Mulvey's formula.

9. While it is beyond the scope of this essay, a number of writers have, in fact, questioned the viability of such approaches for film analysis itself. Scott McQuire, for example, has expressed concern over the way critics and theorists have prioritized thematic readings that obscure any sense of difference in examples of horror. Film examples are mapped out through ahistorical approaches; the methodology consists of "mapping the surface elements of different discrete texts onto an underlying meta-narrative which is then presented as the real meaning of the various symbolic and iconographic devices" (McQuire 1987, 24). Eileen Meehan has similarly stated that these kinds of cause-and-effect approaches require "an assumptive leap that reduces consciousness, culture, and media to reflections of each other" (1991, 48). In many ways, computer games highlight a problem already inherent in film theory and criticism; and the inadequacy of this form of criticism becomes even more overt when one applies similar interpretations to computer games.

10. Landow has suggested that hypertext theory parallels poststructuralism and other theoretical traditions in its concerns with the open text—suggesting Kristeva on intertextuality, Bakhtin on dialogism and multivocality, Foucault on networks of power, and Deleuze and Guattari on rhizomatic and nomadic thought (1994, 1).

WORKS CITED

Aarseth, Espen J. 1994. Nonlinearity and Literary Theory. In *Hyper/Text/Theory*, ed. George P. Landow, 51–86. Baltimore: Johns Hopkins University Press.

Barnow, Eric. 1981. *The Magician and the Cinema*. Oxford: Oxford University Press.

Bordwell, David. 1989. *Making Meaning: Inference and Rhetoric in the Interpretation of Cinema*. Cambridge: Harvard University Press.

Carroll, Noël. 1988. *Mystifying Movies: Fads and Fallacies in Contemporary Film Theory*. New York: Columbia University Press.

Clover, Carol. 1992. *Men, Women and Chainsaws: Gender in the Modern Horror Film*. Princeton: Princeton University Press.

Collins, Jim. 1991. Batman: The Movie, Narrative: The Hyperconscious. In *The Many Lives of the Batman: Critical Approaches to a Superhero and His Media*, ed. Roberta E. Pearson and William Uricchio, 164–81. New York: Routledge.

Dika, Vera. 1990. *Games of Terror: Halloween, Friday 13th, and the Films of the Stalker Cycle*. Toronto: Associated University Press.

Docker, John. 1994. *Postmodernism and Popular Culture: A Cultural History*. Cambridge: Cambridge University Press.

Ferrell, Keith. 1994. Interactive Storytelling. *Electronic Entertainment*, August, 30.

Foster, Kirsten. 1995. System Shockers. *PC Power* 13 (January): 32–35.

Gunning, Tom. 1995. Phantom Images and Modern Manifestations: Spirit Photography, Magic Theater, Trick Films, and Photography's Uncanny. In *Fugitive Images: From Photography to Video*, ed. Patrice Petro. Bloomington: Indiana University Press.

Heard, Mervyn. 1996. The Magic Lantern's Wild Years. In *Cinema: The Beginnings and the Future*, ed. Christopher Williams, 24–32. London: University of Westminster Press.

Landow, George P. 1994. "What's a Critic to Do?": Critical Theory in the Age of the Hypertext. In *Hyper/Text/Theory*, ed. George P. Landow, 1–48. Baltimore: Johns Hopkins University Press.

Liestol, Gunnar. 1994. Wittgenstein, Genette, and the Reader's Narrative in Hypertext. In *Hyper/Text/Theory*, ed. George P. Landow, 87–120. Baltimore: Johns Hopkins University Press.

McGrath, Roberta. 1996. Natural Magic and Science Fiction: Instruction, Amusement and the Popular Show, 1795–1895. In *Cinema: The Beginnings and the Future*, ed. Christopher Williams, 13–23. London: University of Westminster Press.

McQuire, Scott. 1987. Horror: Re-Makes and Offspring. *Antithesis* 1, no. 1: 21–25.

Mayne, Judith. 1993. *Cinema and Spectatorship*. London: Routledge.

Meehan, Eileen E. 1991. "Holy Commodity Fetish, Batman!": The Political Economy of a Commercial Intertext. In *The Many Lives of the Batman: Critical Approaches to a Superhero and His Media*, ed. Roberta E. Pearson and William Uricchio, 47–65. New York: Routledge.

Mulvey, Laura. 1975. Visual Pleasure and Narrative Cinema. *Screen* 16:6–18.

———. 1981. Afterthoughts on "Visual Pleasure and Narrative Cinema" Inspired by King Vidor's *Duel in the Sun*. *Framework* 15–17:12–15.

Paul, William. 1994. *Laughing Screaming: Modern Hollywood Horror and Comedy*. New York: Columbia University Press.

Pearson, Roberta E., and William Uricchio. 1991. Introduction to *The Many Lives of the Batman: Critical Approaches to a Superhero and His Media*, ed. Roberta E. Pearson and William Uricchio, 1–3. New York: Routledge.

Penn, Gary. 1994. Rendered Dull and Devoid. *PC Gamer* 1, no. 10 (September): 86.

Snyder, Ilna. 1996. *Hypertext: The Electronic Labyrinth*. Melbourne: Melbourne University Press.

Chapter Five

Busy Box Interface
The Pleasures of Winding

Brian Kelly with Scott Bukatman

I'm a boogie-woogie baby
if you wanna see me boogie
all you gotta do is wind me up.
—Bootsy Collins, "Wind Me Up"

Digital Machines: The Return of the Mechanical

On the monitor the infinitely mobile camera, a simulated kino-eye (Michelson, 1984) whose movement is constrained only by the limits of the director's imagination and the agility of human vision, carves a roller-coasting path through the invisible Cartesian grid that defines our relative position in a three-dimensional frame that feels mostly frameless. A spinning, self-acting, heavy metallic cabinet comes to the fore accompanied by an industrialized Philip Glass–type soundtrack. The doors fling open. The deployment of robotic armatures lights the way to an electronic display of psychedelic test patterns that would have delighted Timothy Leary. The penetration of that screen reveals an otherworldly terrestrial sphere ensconced in the scarlet atmosphere of Ray Bradbury's Mars. The barren landscape is home to an urban oasis, a Martian Manhattan, a glittering city-machine protected by an enormous, streamlined variant of Buckminster Fuller's geodesic dome. We come to rest at the foot of this municipal iron lung and are confronted by an agglomeration of massive, stainless steel letters that

Fig. 5.1. Opening shot of *L-Zone*. Director Haruhiko Shono's domed and abandoned digital city plays like a hands-on museum of industrial artifacts.

rise from the cracked desert surface to announce the title and main character of this CD-ROM: *L-Zone* (1993).

The opening sequence of *L-Zone* relies heavily on our intertextual familiarity with the genre of science fiction to position us within the tensions generated by the blending of old and new technological forms. Its aesthetic strangeness and transparent but unambiguous separation of inside and out inspire a childlike curiosity. The *L*, marked by a variety of industrial accoutrements, flashing lights, and galvanized stairs, magnetizes our vision and compels investigation. Programmed by both instinct and culture, we click on it. A stuttering series of shots brings us face to face with a rectangular portal. The door slides crisply open on push-button cue, the sound and feel of it easy and familiar to *Star Trek* fans. Beyond it only darkness. Animated by the lure of the unknown, we proceed eagerly: an airlock—another door—darkness—a shuttle train terminal—a shuttle train; an eerie recombination of the virtuality of Max Headroom and the drugged imaginings of Jim Morrison. Not all the clicks work. Spaces have to be searched, clues have to be found, machines have to be activated.

Devoid of narrative meaning and the pursuit of victory, *L-Zone* claims no teleology. We may eventually stumble onto planet Green as in the final scene of *Blade Runner*, but the show is about the machines, not the garden (*L'machine pour l'machine*). These machines—everything from antique music boxes and welding torches to remote-controlled prosthetics and holograms—are set in motion through a bewildering but satisfying array of push buttons, toggle switches, and indicator lights. They are not unlike the busy box suspended over the child's crib; there is no point to their activation, except of course sensory stimulation. We can see, and hear, the machines move, squeak, and groan, and somehow it *feels* as though we can get our hands on them. The phenomenology of tactility that emerges is made possible by combining vision with (digitally reproduced) mechanical sound. The phenomenological weight of the latter, best exemplified by the tensile springs of John Lasseter's endearing and brilliantly anthropomorphized luxor lamps, pushes haptic vision even closer to the sense of touch.

An electronic self-simulation, *L-Zone* calls our attention explicitly to the human-machine interface. In so doing, it exploits the unique capabilities of the medium to demonstrate and restore some of what's been lost in the transition from mechanical participation to electronic separation. It does this by providing imagery that is explicitly mechanical in its appeal and design, but is, in actuality, fully digital. Its obsessively industrial aesthetic extends and supports a wider circulation of print ads, televisual graphics designs, and cinematic imagery recalling machine age materials and technologies: riveted steel plates, cast-iron gears and girders, trains, electricity, and skyscrapers. To the extent that recollections of this sort can be said to speak directly to the present, we suspect that industrialism has returned from the repressed not merely as quaint aesthetic pleasure (a kind of Rubik's Cube under Tiffany glass), but as a previously forgotten context for making *sense* of the electronic interface. It is about rediscovering the mechanicity of digitization, which, in addition to counterbalancing the discourses of disappearance, shifts our understanding of the character of *digital* memory—which now must be considered in relation to *industrial* memory.

Industrial Memory: Wind Me Up, Wind Me Down

Perhaps anticipating the pending conversion of Bethlehem Steel, *L-Zone* auteur Haruhiko Shono's teaser reads like public relations copy for a living, "hands-on" museum of science and industry: "A hyper-automated city built by a mad scientist . . . it is a realm that has with the passage of time begun slowly to deteriorate. . . . The banks of machines and dynamos wait for some agency to switch them on and so awaken *L-Zone* to function as it did in the past" (Shono, 1993, 10). The irreversibly scorched and outmoded lungs of Bethlehem Steel, that industrial behemoth straddling the Lehigh River in the Appalachian foothills of eastern Pennsylvania, are today gasping for air with hopeless functional and financial inefficiency. Doomed by its expanding obsolescence, this nineteenth-century giant of city-factories sinks slowly into decrepitude. Its once awesome production capacity spent, the hulking machinery chokes and sputters as if nearing its death. Anticipating the inevitable, the Smithsonian stands watch, its highly skilled team of museum officials determined to embalm the "original" and memorialize the sight/site for posterity.

The planned conversion of the gargantuan Bethlehem Steel compound from vast junkyard to museum and tourist attraction can be read as perhaps the grandest gesture to date made on behalf of what Mark Dery calls "industrial memory." The drive to preserve the facility's monumental physical presence is testimony to what he calls the "recrudescence of the mechanical paradigm" (1996). As the manufacturers and durable goods of the industrial economy give way to the symbolic analysts and immaterial commodities of the information age, the old machines are pushed from the exigencies of everyday life into the realm of culture. The perception of idling, obsolete factories subsequently shifts from a market to an aesthetic axis. And though machines—their brute physicality, contoured surfaces, and rhythmically moving parts—have long made terrifyingly terrific visual spectacle, their postindustrial "absence" renews the urgency of their cultural presence.[1]

Cinematic re-presentations of machine age machinery, such as *Ballet Mécanique* (1924), helped assimilate those machines to the terms of human vision. Late-model mechanical tropes are also motivated by anxieties energized by the destabilization of modern sociocultural forms generally, and transcendentalist imaginations dreaming of

Fig. 5.2. Mechanical tropes revive industrial memory and complicate electronic discourses of disappearance.

downloaded immortalities—disembodied, gravity-free, silicon-based, posthuman life forms—more specifically. The dream of interface evolutionists, exemplified by Jean Baudrillard's (1983) fractal subject, Hans Moravec's (1988) mind children, and Andre Bazin's (1992a, b) myth of total cinema, is to do away with all mediating artifacts entirely; the production of a seamless Chardinian "noosphere" devoid of machines, language, and even bodies that might introduce friction and obstruct the achievement of the undistorted communication envisioned by Jürgen Habermas (1983). Nicholas Negroponte writes, "Therein lies the secret to interface design: make it go away" (1995, 93). Such is the mantra of what Don Ihde has called a "doubled desire: on one side, . . . a wish for total transparency, total embodiment, for the technology to truly 'become me' . . . [on] the other, the desire to have the power, the transformation that the technology makes available." This contradictory desire "both secretly *rejects* what technologies are and overlooks the transformational effects which are necessarily tied to human-technology relations" (1983, 75).

The current state of electronic interfaces is such that the surreal, instantaneous movement from *here* to *there* eludes phenomenological experience. Borrowing from cinematic montage, interface displays enact, at the touch of a button, not so much a journey, but a discrete succession of being-theres, a privileging of the synchronic over the diachronic, what Douglas Kellner calls an "acceleration of inertia" (1994, 12). Not surprisingly then, much has been said about the spreading of a generalized anaesthesia among us postwar moderns.[2] Marshall McLuhan, for one, writes that the "effect of electric technology had at first been anxiety. . . . Now it appears to create boredom" (1964, 26). The experience of electronic difference is created not through the temporal interplay of bodily and mechanical forces, but through a relentless visual juxtaposition of speed-of-light imagery. Subsequent to an initial hysteria, "narcosis" and "mesmerization" (Baudrillard's metaphors) more accurately describe channel- or Web-surfing than does "drama."

Discursive treatments, however, of what Fredric Jameson (1991) calls "the waning of affect" do little to explain the manifold stocks of cultural phenomena that reflect the ceaseless Heideggerian gathering and dispersion of energies; the windups and wind-downs so well embodied by winding toys.[3] From popular narrative cinema, with its endless fireballs of apocalyptic desperation, to major team sporting events like baseball ("The windup, the pitch . . .") and American football, the society of the www.spectacle.com offers its participant-observers repeated cycles of perceptual/emotional flows of intensity in which invested energies are alternately compressed and exploded. Forms vary enormously between contexts, but the pattern is unmistakable: sexual pleasures; the ebb and flow design of games of all kinds; modern work schedules including happy hour martinis and prime-time television; shopping for and finding oneself in commodities; countdowns, deadlines, and buzzer-beaters.

(D)Evolution of the Interface: Restoring Tactility

For the past year, NBC has been airing a series of self-promotional ads in which its longtime logo is brought to life through some rather unexpected media. One such animation involves the brief but mesmerizing performance of a vibrantly colored windup toy. Emerging stage right with the smooth confidence of a locomotive, the spotlighted peacock quickly loses speed, and finally its balance, as the energy stored in its internal gear structure presumably dissipates below threshold. The derailed toy ejects its key before finally succumbing to the forces of inertia. Though the toy is immediately morphed into the abstractly generic peacock that guarantees network recognition, the afterimage is decidedly nonelectronic. That is, the windup process itself goes unrepresented, but is suggested by the key, whose presence marks the toy as unmistakably mechanical and forecloses questions about how the toy is animated. A textbook example of McLuhan's famous aphorism, this toy, electronically modeled as a work of mechanized art, foregrounds its mechanicity while hiding its digital configuration.[4] Unlike its series of self-reflexive ads featuring prime-time sitcom actors playfully wielding remotes the size of stone tablets, NBC's mechanical peacock shifts our attention from pushbutton electronics to the mechanical cranks that are forever disappearing in the rearview mirrors of memory and museums.

Since the machine entered the garden of Jefferson's yeoman farmers, technology has been extolled for its promise of the effortless life. Machines disburdened the body of strenuous physical toil and extended the range of its influence. Machine age machines, however, from telephones and phonographs to automobiles and airplanes, were far from today's self-acting varieties so famously described by Donna Haraway (1985), and so well animated in *The Jetsons*. Many of these mechanical relics were instead designed for activation by crank-handle. Depending on the size of the crank, the engines of these machines could be wound with everything from a few turns of the arm to full bodily engagement. Not every aspiring pilot, however, could crank up a single-engine aircraft as handsomely as a bare-chested Burt Lancaster.[5] Silent comedy was filled with emasculated husbands, their hand-cranked flivvers running out of "juice" while angry cops and righteous mothers-in-law looked on in wrathful judgment.

Cranks demanded a more democratic and efficient alternative. They have since become keys, which more recently have been transformed into slide-through keycards and remote activators. A trivial example perhaps, but convincing evidence nonetheless of a larger trend in which knowledge of the interior workings of our most familiar machines is gradually moving from the somatic to the symbolic.[6] Such is the case with the tiny, whirring *ballet mécanique* set in unseen motion by the button that activates your CD-ROM. All that remains of the pulleys and hoists of the block and tackle, one of the earliest forms of servomechanism, is its space age abstraction: the block-and-triangle icon.

The actual operational means of sophisticated machinery tend to remain mysterious in any age, but unlike users of digital technologies or better computerized machines, which reveal nothing of their interior machinations or capabilities vis-à-vis the human animal, ordinary users of strictly mechanical systems were able to sense, through their literal flesh, something of the machine's internal dynamic and structural limitations. This is not to deny N. Katherine Hayles's "materiality of informatics" (1993), but to point to an increased recession of body *and* machine in the shift from cranks to buttons as interface practices. Having reached its Ellulian (1964) apotheosis at the second millennium, electronic efficiency precludes any muscular memory of machines. The meaning of the block-and-triangle iconography that aspires to universal decipherability instead implodes into semiotic oblivion.

This withdrawal, or absent presence of the machine, coupled with the always absent-present body, exacerbates anxieties associated with digital etherealization.[7] Compelled in part by the rhetorical excesses of postmodernism, we chase our bodies into the spaces of the machine, into computers, into cyberspace, to get a *sense* of what these machines are about, to find out if there is knowledge in there (Baudrillard), to imagine what cannot be seen (Thrift), and to reassure ourselves of the continued viability of our bodies (Bukatman). Traditional media rely heavily on a phenomenology of vision to recall bodies and machines, but we also crave a phenomenology of tactility to re-member or re-attach them. It is no longer enough to watch life live itself through the digital windup toy—we want to experience the pleasures of winding it.

Although the windup toy enjoys a newfound celebrity status as both collectible and feature film star, its digital versions fail to fully restore its rambunctious mechanical delights. Evoking what Freud might have called the primal mechanical scene, windup toys, unlike animation by imagination or handy manipulation, involve the transfer of muscular energy (capacity for work) across the human-machine interface. Through what Merleau-Ponty (1962) described as *flesh*, the winding agent is able to explore the limits of the toy's ability to maintain tension. Information (that is, *matter-energy*, as in Claude

Shannon's technical model)[8] is reciprocally and invisibly passed between the organic body of the agent and the manufactured body of the machine-toy. Though the specific processes of the mechanical translation and storage of energy remain mysteriously hidden beneath the toy's exterior, the distancing and suspension of that energy serve only to heighten the anticipation of its release.

Like mainstream narrative film (which must also be wound before it can roll)—or life itself—the ending is known, but the path and its duration are not. The pleasure of visually chasing the toy is tied directly to our memory of physically imparting our own work to what once was a lifeless object. That memory is produced over a sensible period of time that permits the playful buildup of anticipation. As a product of the Imaginary (i.e., the imagination of a future state), anticipation is, in effect, the uncertain extrapolation of the memory of a previous state. Mirrored in the resistance of the internal mechanism of the toy, anticipation can be savored and torqued to the brink of a Sadean materiality. Anticipation of discharge, rooted in the *material* memory of the movement from diffusion to concentration, adds a phenomenology of tactility to the phenomenology of vision that produces its pleasures.

Whereas we imagine ourselves, our energy, manifested in the toy's frolicking movements, Debord's (1983) spectator simply cannot make a similarly strong identification. Vicariously disembodied spectatorial pleasures have no remembrance of the work that ultimately produces the drama. Spectators may enjoy the feeling of being wound, but they have no recourse to the pleasures attained from turning the key and loading the spring. The difference might best be explained metaphorically. Think of two children fighting over a toy, or their tortured anticipation while awaiting their turn at the controls of a video or computer game. Think of how eager they are to get their hands on things. Think of almost any Nickelodeon commercial, or the pleasure people take in making things "work."

CD-ROM: The Future (Imagination) of the Interface

CD-ROM, a nascent (and perhaps soon-to-be-obsolete) medium, the product of a particular historical and technological convergence, puts in the hands of the player a scope of control previously unknown to related media. Least like film, more like television, and more still like Nintendo, CD-ROM technology provides its users significantly more agency with regard to the duration and sequence of onscreen events. Although not as immersive as Omnimax and virtual reality—which effectively demolish *off*screen space—CD-ROM (not unlike painting) adds two degrees of control to the subjective performance of vision: how long to dwell in a space and where to go next, decisions traditionally made exclusively by directors (VCRs and remote controls notwithstanding). Rather than surrendering to a director's vision of a particular narrative-visual-psychological journey, we are obliged to *work* our way through the spatiality of the image, making CD-ROM spaces tactile beyond haptic vision. Like the urban metropolis whose mazes are etched in stone, the read-only-memory of the disc lies in wait. It can be activated only by the adventuring cyber-*flaneur*.

Gadget: *Invention, Travel, & Adventure* (1994), like *L-Zone*, employs two visual aspects: the interactive position that provides "tactile" access to its digital environments, and the cinematic perspective that returns the experience to a phenomenology of vision. We are asked to collect gadgets and clues as we navigate a labyrinth of virtual spaces. Unlike *L-Zone*, *Gadget* unfolds as a narrative in which digital objects take on contextual meaning. It includes a cast of "human" characters; refugees, it seems, from Monty Python's animation files. Reminiscent of Terry Gilliam's *Brazil*, the obsessive interiority of the interactive environments (hotel rooms, lobbies, train stations, train cars and private rooms, museums, etc.) are balanced against the exteriors of the cinematic sequences, which are rendered almost exclusively in black and white.

The story unwraps as a corrupt military conspiracy involving high-placed megalomaniacal scientists and government officials who cloak their experiments in an insidiously devised "save the world" (from an approaching meteorite) campaign. Those experiments relate to the successful development of a memory machine called the Sensorama. We, in fact, are the thirteenth subject to be tested. Along the way, before the "unexpected truth is revealed," we are fed a steady dose of lies by coconspirators and confused by the incoherent babblings of previous subjects. Says one such unfortunate participant, "Every time the whistle sounds, a grating noise creeps from deep in my ears up the back of my skull. I need a corset made of iron to support my head, a thick steel collar that'll stand up to superheated steam" (Shono, 1994).

Like Proust's (1981) Marcel, we awake in a hotel room we have absolutely no recollection of; we are not quite sure who we are. We begin looking for clues, visually identifying familiar objects, moving around the space, searching its boundaries for differences, objects, doors, exits. We gather information, and with it, memory. That information is of course already stored in the memory of the machine, the computer. The process of moving that memory from the machine to the player has everything to do with the interface. Because the CD-ROM provides its players with subjective control over time and direction of movement, the computer must faithfully reproduce the space regardless of how we decide to negotiate it. To use a more appropriate metaphor, the space must be bolted down in the memory of the machine. And, barring technical breakdown or user illiteracy, it performs in precise accordance with its promise: immediate and error-free recall.

The computer is represented in *Gadget* by the Sensorama, an oversized assemblage of vacuum tubes protected by an oversized industrial light fixture. Its hypnotizing patterns of light, movement, and sound have the effect of rendering the subject's memory entirely controllable, a clear reference to the anxieties of digital disappearance (including those related to the politics and power dynamics of its rhetorics and [virtual] realities). What the Sensorama finally secures for its makers, "absolute memory control," becomes for Shono an occasion to sneer at the hubris and impossibility of its ambition. At the same time, he makes marvelous use of its latest technological expression, the digital computer itself, to make his point.

Human memory is less reproducible. We select, distort, and misremember memories, which are produced, in the first instance, through the body, through the Imaginary. They are forever subject to revision and contestation. That is, the accumulated memories we bring to the game gather to make sense of what we see there. New perceptions, in turn, alter those memories, which are then projected through the Imaginary to anticipate what might be next. In this way, our memory becomes incessantly mutable, in constant tension between past, present, and future.

The Imaginary is symbolized in *Gadget* by the boy who floats in and out of the narrative like the memories of childhood (past), as well as by the image-premonitions (future) that turn up on various monitors throughout. It is also evoked by the cinematic sequences themselves (see Metz, 1982), as it is in *The City of Lost Children*, the CD-ROM based on the film of the same title. In *City*, however, the Imaginary is less represented than it is *exercised*. Not only are we asked to locate the pieces of the puzzle, we must deconstruct their familiar meanings and imagine their possibilities in unfamiliar contexts: a metal bar must be used to short-circuit the lighthouse control panel; a wire brush jams open and disables the cash register that maintains an electrical security system; a candle provides the means to sever a rope slowly enough to avoid the disastrous consequences of cutting it.

Exploring the potentiality of things requires an active forgetting of their habitual uses. Their meanings must be crossed out, constantly put under erasure. Signifiers must be wrenched loose from signifieds, and set adrift on the sea of Derrida's (1970) endless chain of nonsignification. But forgetting is not as simple a matter as deleting files from the hard drive. Generally speaking, it is a task made more difficult with the passage of time. In *City*, we are not positioned kino-subjectively in the manner of cinema (except of course for the cinematic sequences), but rather in the place of the electronic puppeteer. Pulling the strings of digital code that control the movements of our stand-in, Miette (a streetwise, twelve-year-old orphan), we can experience the advantage of a child's weaker cultural attachments. Because their investments in meaning are less serious, children can more easily dismember signs and imagine the multiplicity of their possible meanings. The re-membering of signifiers and signifieds becomes a creatively synthetic activity through which new forms of meaning emerge. To Miette, paint is more than a cosmetic application of color, it also means the obstruction of vision.

New forms, whether spontaneously self-organized or the product of years of research and development, cannot be contained by the imaginations that give them shape. The mad scientist, be it Baron von Frankenstein, the brain-in-a-vat Irvin in *City*, or the sadistic Sid of *Toy Story* fame, learns quickly that "living things can turn against their creators" (Asendorf, 1993, 191). Sid's genius for mechanical bricolage starts with an impulse to deconstruct the already-made, which, in excess, becomes the urge to obliterate. In a childhood version of the Slotkinian (1973) myth of "regeneration through violence," Sid tortures and annihilates his mutant toys with unbridled delight. Paralyzed by fear, they wait to see which of their numbers will be his next victim. That is, until those all-American frontiersmen, Woody and Buzz, the cowboy and astronaut respectively, accidentally venture beyond their boundaries and into the spaces of Sid's darkness.

Woody and Buzz can be read as technology past and future. Woody's six-shooter, the spring of which must be loaded mechanically, stands in stark contrast to the hi-tech laser mounted on Buzz's wrist. The tensions between them, motivated primarily by Woody's fears of obsolescence, reflect the tensions generated by the move from mechanical to electronic interfaces. In *Toy Story*, those tensions lead to the realm of Sid's imagination, which, when viewed through a magnifying glass, reveals its obsession: the re-membering of body and machine. In its etymological sense, to *re-member* is to re-*flesh*; to reattach to the body what had been dismembered; to bring to consciousness (i.e., the screen, the interface) that which has been forgotten, relegated to memory. And who better than an adolescent with braces to recall the materiality and signifying power of metal and flesh?

Sid's recombinatory operations on humanoid and mechanical toys bring the technological interface into high relief. Nearly all his reconstructions involve the radical re-attachment of body and machine: the one-eyed, stubble-haired baby doll head on mechanical spider legs; the hand-in-the-box; the fishing rod joined at the hip to human legs; the human torso on skateboard; etc. These creatures graphically depict what Haraway explains as the implosion of categorical certainties related to humans, animals, and machines; the promises of Sid's monsters lie in the recognition that the interface is always already *a sign of difference*.

We have already noted that in comparison to the electronic, the mechanical interface demands greater bodily engagement, takes longer to actuate, and consequently involves a more sensible journey that registers analogically in somatic memory and symbolic consciousness. The electronic abstraction of those processes shifts the interface experience from the sensible energy transfer between the internal anatomies—the meat and gears—of humans and machines to an interplay of selected external sensory organs: the eyes and hands of the former and the push buttons and displays of the latter. The popular alien figure drawn from the description provided by author Whitley Strieber, and shamelessly exploited by Roswell, New Mexico, uncannily reflects the new interface demands: oversized, darkened eyes; long, slender fingers; atrophied body and body parts—ears, nose, mouth (just big enough for guzzling Surge).

Digital difference, in other words, is less a matter of travel, tension-building, and memory than of being here, now there, and now everywhere and nowhere. Industrial memory is not so much about the past as about the present, insofar as the industrial aesthetic makes explicitly present what is absent-present in most electronic interfaces: a physical sense of energy transfer and storage, the sensible experience of time, and, following that, movement. Differences conditioning the very possibility of movement are defined by the boundaries, "real" or imagined, that make sense of those differences. If differences, boundaries, interfaces cannot be sensed, movement becomes stasis, and the disappearance of the interface begins to look more dystopian than utopian. When movement itself is superseded by the quasi-transcendental instantaneity of microelectronics, there is no memory of moving, no perceptual or sensational registering of the connection between now and then, between signifiers, between self and world. The interface evaporated, all differences effaced, theorizing begins on the end of history, meaning, and the body.

Re-membering the Memory Machine

On the island of Manhattan, less than a hundred miles to the east of Bethlehem Steel, Gary Kasparov, reigning world chess champion, resigns his match with Deep Blue, IBM's latest chess-playing *übermaschine*. Kasparov's defeat marks the first time in the history of computing that a calculating machine has outwitted its human counterpart. Alarmists and technophobes brand the surrender a telling metaphor, a sure sign of dystopian things to come. Talking heads, confidently insisting that the human is still in control of The Machine, remind us that HAL (of *2001: A Space Odyssey*) is a purely fictional creation, not a fateful inevitability. The Smithsonian team, its keen cultural nose perhaps sniffing out the most decisive turn yet toward a postbiological future, doubtless anticipates preserving the event, if not Man, for the sake of posterity.

The astonished Kasparov's failure to triumph over Deep Blue adds another layer of reverence for the computer that has until now been withheld for just this kind of peculiar fin-de-millennium science fictive moment. It is a reverence grounded in the struggle to comprehend what it means for a seemingly inert machine to consider two hundred million possible chess moves in a single second. A kind of technomysticism ensues as metaphors are proposed to render the event sensible: How does Deep Blue *do* that? If, as John Clute suggests, its " 'actions' are invisible, then (perhaps) our fates are likewise beyond our grasp."[9] Lacking moving parts and aesthetic charisma, the quintessential "black box" resists superficial representation, compounding the problem of its knowability in a culture so heavily invested in the style and performance of vision.[10] Dery writes, "sealed in a smooth, inscrutable shell, the computer's inner workings are too complex, too changeable for the imagination to gain purchase on them; only when it is imaged in the heavy metal of the Machine Age can this post-industrial engine be grasped" (1996).

Dery's metaphors are telling. Because cyberspace is forever inaccessible to the body, the seat of sensation, perception, and consciousness, the best we can do is imagine/image it in ways that make it "graspable." In *City*, for example, we can, through Miette, walk, talk, and grasp objects. But the computer is not a material extension of the body in the manner of mechanical technologies. It is rather an extension, or, following the Krokers, an exteriorization of memory (Kroker and Kroker, 1987). How then can we imagine a graspable memory? In short, we already have. The computer itself is nothing less than the material product of the imagination of memory as bits, lots of zeroes and ones. And, to varying degrees, those bits are also manipulable. Why then do we insist on industrializing the spaces of digital memory?

A digital *tabula rasa*, the computer, like the replicants in *Blade Runner*, has no memory outside its implanted ones (and zeroes). And like Krank in *City*, it has no dreams, no Imaginary. It can perform memory only in accordance with the Symbolic logic mastered by Bill Gates and guided by Negroponte's Media Lab. Despite attempts to transcend the Symbolic in the name of artificial "intelligence," success remains elusive (expert systems and parallel processing designs [Deep Blue] notwithstanding). Bit memory, in other words, behaves mechanistically—literally, done as if by machine, uninfluenced by the Imaginary. That is, like so many automatons, bits function only within the parameters of their programmed instructions.[11] The industrial aesthetic is, in this way, about *imagining the computer's memory as a mechanical memory;* infinitely faster, smaller, and more flexible than its electromechanical forebears to be sure, but distinctly mechanical nonetheless. To image the computer's materiality is to re-cognize it as a fully controllable technology that must bow to its human creators no matter how many matches Deep Blue takes from Kasparov. As with any machine with an astronomical number of moving parts, the computer's tolerances are infinitesimally small. And despite its gigantic capacity to "handle" zeroes and ones, it remains trapped in the Symbolic, unable to generate its own Imaginary, its own meaning, its own future. At least for now . . .

Shortly before his recent death, Carl Sagan mused on the longstanding, hackneyed concern regarding the formidable and multiplying powers of science and technology. Because those powers are neither well nor widely understood, Sagan publicly fretted their apocalyptic potential. To the extent that his position seriously and accurately reflects the late twentieth-century nexus of scientific, technological, and epistemological moments, shifting interface practices may be said to be among those developments that have fostered the crisis he perceived. While it is true that mechanical technologies have always extended and amplified the capacities of the human subject, those capacities have exploded in the atomic age. After all, as in so many science fiction stories, the push of a button could just as easily destroy the planet as it could open a CD-ROM.[12]

NOTES

The authors would like to thank Christie Kelly for her skill, insight, and keen sense of humor at the interface.

1. See, for example, the work of machine age filmmakers like Dziga Vertov and Fritz Lang.

2. A short list includes Herbert Marcuse's (1964) "one-dimensional man," Spencer Weart's (1988) "nuclear fear," Langdon Winner's (1986) technological somnambulism, Susan Buck-Morss's (1992) synaesthetic system, Fredric Jameson's (1991) waning of affect, and Jean-Francois Lyotard's (1984) high modernist *eclipse* of affect.

3. See Heidegger (1977) for a more complete discussion of his notion of "gathering." Also it is worth remembering that numerous studies have shown

that children exhibit hyperactive behavior after watching television for any extended period.

4. In *Understanding Media* (1964), McLuhan argues that the content of any new medium is the old medium.

5. We are claiming artistic license here. Lancaster is inspiring in John Frankenheimer's film *The Train* (1964), though he doesn't actually crank up a single-engine aircraft.

6. It is not our intent here to reify mind-body dualisms.

7. See Ihde (1983) and Turner (1996) for a complete discussion of absence-presence and the body.

8. See Rogers (1962) for an explanation of Shannon's technical model of communication.

9. John Clute quoted in Bukatman (1993, 201).

10. In an ad for the 1997 NBA finals, NBC humorously and effectively demonstrates as much by matching Deep Blue against a highly mobile human body. The former, resembling an overgrown filing cabinet, stands helplessly nearby as a lithe, superbly conditioned athlete makes a spectacular dunking mockery of its immobility. Looking on from the sideline, a tripartite panel of Blue's scientist-keepers can only bemoan its lack of arms.

11. Lauren Wiener (1995) might not agree with this position. Software may be capable of strange behavior, especially in extremely complex applications where code is constantly being added, written over, and so forth. Our point is simply that it is reassuring to think of the computer and its functions as mechanical and hence knowable and controllable.

12. Clearly, as Carol Lee Sanchez (1992) reminds us, the planet will survive the tenure of the human species. The "planet" is us.

WORKS CITED

Asendorf, C. 1993. *Batteries of Life: On the History of Things and Their Perception in Modernity.* Berkeley: University of California Press.

Baudrillard, J. 1983. The Ecstasy of Communication. In *The Anti-Aesthetic: Essays on Postmodern Culture,* ed. H. Foster. Port Townsend, WA: Bay Press.

———. 1989. *America.* London: Verso.

Bazin, A. 1992a. The Myth of Total Cinema. In *Film Theory and Criticism: Introductory Readings,* 4th ed., ed. G. Mast, M. Cohen, and L. Braudy. New York: Oxford University Press.

———. 1992b. The Evolution of the Language of Cinema. In *Film Theory and Criticism: Introductory Readings,* 4th ed., ed. G. Mast, M. Cohen, and L. Braudy. New York: Oxford University Press.

Buck-Morss, S. 1992. Aesthetics and Anaesthetics: Walter Benjamin's Artwork Essay Reconsidered. *October* 62 (fall).
Bukatman, S. 1998. Taking Shape: Morphing and the Performance of Self. Unpublished manuscript.
———. 1993. *Terminal Identity: The Virtual Subject in Postmodern Science Fiction.* Durham: Duke University Press.
Debord, G. 1983. *Society of the Spectacle.* Detroit: Black and Red.
Derrida, J. 1970. Structure, Sign, and Play in the Discourse of the Human Sciences. In *The Structuralist Controversy: The Languages of Criticism and the Sciences of Man,* ed. R. Macksey and E. Donato. Baltimore: Johns Hopkins University Press.
Dery, M. 1996. Industrial Memory. *21C.* http://www.21c.worldideas.com.
Ellul, J. 1964. *The Technological Society.* New York: Knopf.
Habermas, J. 1983. Modernity—An Incomplete Project. In *The Anti-Aesthetic: Essays on Postmodern Culture,* ed. H. Foster. Port Townsend, WA: Bay Press.
Haraway, D. 1985. A Manifesto for Cyborgs: Science, Technology, and Socialist Feminism in the 1980s. *Socialist Review* 15, no. 2 (March–April).
———. 1991. *Simians, Cyborgs, and Women: The Reinvention of Nature.* New York: Routledge.
Hayles, N. K. 1993. The Materiality of Informatics. *Configurations* 1 (winter).
Heidegger, M. 1977. *The Question Concerning Technology and Other Essays.* New York: Harper and Row.
Ihde, D. 1983. *Existential Technics.* Albany: State University of New York Press.
Jameson, F. 1991. *Postmodernism, or The Cultural Logic of Late Capitalism.* Durham: Duke University Press.
Kellner, D., ed. 1994. *Baudrillard: A Critical Reader.* Cambridge: Blackwell.
Kroker, A., and M. Kroker. 1987. Body Digest: Theses on the Disappearing Body in the Hyper-Modern Condition. *Canadian Journal of Political and Social Theory* 11, nos. 1–2.
Lyotard, J. 1984. *The Postmodern Condition: A Report on Knowledge.* Trans. G. Bennington and B. Massumi. Minneapolis: University of Minnesota Press.
Marcuse, H. 1964. *One-Dimensional Man.* Boston: Beacon Press.
McLuhan, M. 1964. *Understanding Media: The Extensions of Man.* New York: McGraw-Hill.
Merleau-Ponty, M. 1962. *Phenomenology of Perception.* Trans. C. Smith. London: Routledge and Kegan Paul.
Metz, C. 1982. *Psychoanalysis and Cinema: The Imaginary Signifier.* Trans. C. Britton, A. Williams, B. Brewster, and A. Gizzetti. London: Macmillan.
Michelson, A., ed. 1984. *Kino-Eye: The Writings of Dziga Vertov.* Berkeley: University of California Press.
Moravec, H. 1988. *Mind Children: The Future of Robot and Human Intelligence.* Cambridge: Harvard University Press.

Negroponte, N. 1995. *Being Digital*. New York: Knopf.
Proust, M. 1981. *Swann's Way*. Trans. C. Moncrieff and T. Gilmartin. New York: Random House.
Rogers, E. 1962. *Diffusion of Innovations*. New York: Free Press.
Sanchez, C. 1992. Animal, Vegetable, Mineral: The Sacred Connection, *Ecofeminism*, fall.
Shono, H. 1993. *L-Zone*. CD-ROM, Japan.
———. 1994. *Gadget: Invention, Travel, & Adventure*. CD-ROM, Japan.
Slotkin, R. 1973. *Regeneration through Violence: The Mythology of the American Frontier, 1600–1860*. Middletown, CT: Wesleyan University Press.
Thrift, N. 1996. *Spatial Formations*. Thousand Oaks, CA: Sage.
Turner, B. 1996. *Body and Society*. Thousand Oaks, CA: Sage.
Weart, S. 1988. *Nuclear Fear: A History of Images*. Cambridge: Harvard University Press.
Wiener, L. 1993. *Digital Woes: Why We Should Not Depend on Software*. Reading, MA: Addison-Wesley.
Winner, L. 1986. *The Whale and the Reactor: A Search for Limits in an Age of High Technology*. Chicago: University of Chicago Press.

Chapter Six

Civilization and Its Discontents
Simulation, Subjectivity, and Space

Ted Friedman

New Paradigms, Old Lessons

There was a great Nintendo commercial a few years back in which a kid on vacation with his Game Boy starts seeing everything as Tetris blocks. Mount Rushmore, the Rockies, the Grand Canyon all morph into rows of squares, just waiting to drop, rotate, and slide into place. The effect is eerie, but familiar to anyone who's ever played the game. The commercial captures the most remarkable quality of video and computer games: the way they seem to restructure perception, so that even after you've stopped playing, you continue to look at the world a little differently.

This phenomenon can be dangerous—as when I finished up a roll of quarters on Pole Position, walked out to my car, and didn't realize for a half mile or so that I was still driving as if I were in a video game, darting past cars and hewing to the inside lane on curves. More subtly, when the world looks like one big video game, it may become easier to lose track of the human consequences of real-life violence and war.

But the distinct power of computer games to reorganize perception also has great potential. Computer games can be powerful tools for communicating not just specific ideas, but structures of thought—whole ways of making sense of the world. Just as Tetris, on the simple level of spatial geometry, encourages you to discover previously unnoticed patterns in the natural landscape, more sophisticated games can teach you how to recognize more complex interrelationships. The simulation game *Sim City*, for example, immerses you in the dynamics of building and developing a city, from zoning neighborhoods to building roads to managing the police force. In learning how to play the game, you develop an intuitive sense of how each aspect of the city affects and is affected by other aspects of the city—how, for example, the development of a single residential area will affect traffic, pollution, crime, and commerce throughout the city. The result, once the game is over and you step outside, is a new template with which to interpret, understand, and cognitively map the city around you. You no longer see your neighborhood in isolation, but as one zone influenced by and influencing the many other zones that make up your town.[1]

Any medium, of course, can teach you how to see life in new ways. When you read a book, in a sense you're learning how to think like the author. And as film theorists have long noted, classical Hollywood narrative teaches viewers not just how to look at a screen, but how to gaze at the world. But for the most part, the opportunities for these media to reorient our perceptions today are limited by their stylistic familiarity. A particularly visionary author or director may occasionally confound our expectations and show us new ways to read or watch. But for the most part the codes of literary and film narrative are set. We may learn new things in a great book or movie, but we almost always encounter them in familiar ways.

Computer games, by contrast, are a new medium, still in flux. While game genres have begun to form, they remain fluid, open-ended. The rules and expectations for computer games are not yet set in stone. Each new game must rethink how it should engage the player, and the best games succeed by discovering new structures of interaction, inventing new genres. What would be avant-garde in film or literature—breaking with familiar forms of representation, developing new modes of address—is standard operating procedure in the world of computer games. Every software developer is always looking for the next "killer application"—the newest paradigm-buster.

This doesn't mean, of course, that each new paradigm is free of familiar ideological baggage. Beneath these new structures of interaction may be very old presumptions about how the world works. *Sim City* may help us see cities with new eyes, but the lessons it teaches us about cities—the political and economic premises it rests on—are conventionally capitalist, if somewhat liberal. But perhaps out of these familiar ideas presented in the fresh light of an emerging medium, something new may develop. At the least, as computer games discover new tools for communicating ways of thinking, new opportunities are opened for more radical visions.

A closer look at the semiotics of a specific computer game can help specify in what ways these new texts teach us to look at the world differently, and in what ways they tell the same old stories. In this essay, I will look at one game that typifies the medium today in all its contradictions. *Civilization II*, a "simulation" game that puts the player in the position of a nation's leader building an empire, radically challenges conventional norms of textual interaction in some ways. Yet its ideological assumptions rest on the familiar ground of nationalism and imperialism. Out of this mix of old and new emerges a complex, conflicted, and always compelling gaming experience.

Booting Up Civilization II

Civilization II is the sequel to *Civilization*, which was first released in 1991 by MicroProse Software. *Civilization II* appeared in 1996. (*Civilization II* adds new features and spices up the graphics from the original *Civilization*, but the basic dynamics of game play remain unchanged. Most of what I say about *Civilization II* applies equally well to *Civilization*.) Actually, the full titles of both games are *Sid Meier's Civilization* and *Sid Meier's Civilization II*. Meier, the cofounder of MicroProse, is the game's inventor and original designer. Meier is known in the computer gaming world for his skill in designing absorbing, detailed simulations. His early games *Pirates* and *Railroad Tycoon* each helped shape the emerging genre in the 1980s. *Civilization* was hailed on its release as one of the greatest computer games ever; *Civilization II* has been similarly honored. In a 1996 survey of the history of computer games, the magazine *Computer Gaming World* named the original *Civilization* the best game of all time ("One Hundred Fifty Best Games" 1996). Rival magazine *PC Gamer* ranked *Civilization II* the fourth best game ever ("Best Ten Games" 1997), and ranked the original *Civilization* one of the fifteen most significant games of all time ("Fifteen Most Significant Games" 1997).

The manual for the original *Civilization* introduces the game this way:

> *Civilization* casts you in the role of the ruler of an entire civilization through many generations, from the founding of the world's first cities 6,000 years in the past to the imminent colonization of space. It combines the forces that shaped history and the evolution of technology in a competitive environment. . . . If you prove an able ruler, your civilization grows larger and even more interesting to manage. Inevitable contact with neighbors opens new doors of opportunity: treaties, embassies, sabotage, trade and war. (Shelley 1991, 7)

What does it feel like to be cast "in the role of ruler of an entire civilization through many generations"? The game follows the conceit that you play the part of a single historical figure. At the beginning of the game you're given a choice of nation and name. From then on, from the wanderings of your first settlers to your final colonization of outer space, the computer will always call you, for example, "Emperor Abraham Lincoln of the United States." Of course, nobody lives for six thousand years, and even the most powerful real-life despots—to say nothing of democratically elected leaders—could never wield the kind of absolute power that *Civilization II* gives even titular presidents and prime ministers. In *Civilization II* you're responsible for directing the military, managing the economy, controlling development in every city of your domain, building wonders of the world, and orchestrating scientific research (with the prescience to know the strategic benefits of each possible discovery, and to schedule accordingly). You make not just the big decisions, but the small ones, too, from deciding where each military unit should move on every turn to choosing which squares of the map grid to develop for resources. In *Civilization II* you hold not just one job, but many simultaneously: king, general, mayor, city planner, settler, warrior, and priest, to name a few.

How does this tangle of roles become the smooth flow of game play? The answer, I think, is that you do not identify with any of these subject positions so much as with *the computer itself*.[2] When you play a simulation game like *Civilization II*, your perspective—the eyes through which you learn to see the game—is not that of any character or set of characters, be they kings, presidents, or even God. The style in which you learn to think doesn't correspond to the way any person usually makes sense of the world. Rather, the pleasures of a simulation game come from inhabiting an unfamiliar, alien mental state: learning to think like a computer.[3]

Cyborg Consciousness

The way computer games teach structures of thought—the way they reorganize perception—is by getting you to internalize the logic of the program. To win, you can't just do whatever you want. You have to figure out what will work within the rules of the game. You must learn to predict the consequences of each move, and anticipate the computer's response. Eventually your decisions become intuitive, as smooth and rapid-fire as the computer's own machinations.

In one sense the computer is your opponent. You have to know how to think like the computer because the computer provides the artificial intelligence that determines the moves of your rival civilizations. Like Kasparov playing Deep Blue, you have to figure out how the computer makes decisions in order to beat it. But in this role of opponent, the computer is only a stand-in for a human player. When multiple players compete, either in *Civilization II* or in the online version, *CivNet*, the AI isn't even needed. And in terms of strategy, the Pentium-powered processor is no Deep Blue; its moves are fairly predictable.

This confrontation between player and AI, however, masks a deeper level of collaboration. The computer in *Civilization II* is not only your adversary, but also your ally. In addition to controlling your rivals, it processes the rules of the game. It tells you when to move, who wins each battle, and how quickly your cities can grow. It responds instantly to your every touch of the mouse, so that when you move your hand along the mousepad, it seems as if you're actually physically moving the pointer on the screen, rather than simply sending digital information to the computer. It runs the universe that you inhabit when you play the game. "Thinking like the computer" means thinking *along with* the computer, becoming an extension of the computer's processes.[4]

This helps explain the strange sense of self-dissolution created by computer games, the way games "suck you in." The pleasure of computer games is in entering into a computer-like mental state: responding as automatically as the computer, processing information as effortlessly, replacing sentient cognition with the blank hum of computation. When a game of *Civilization II* really gets rolling, the decisions are effortless, instantaneous, chosen without self-conscious thought. The result is an almost meditative state, in which you aren't just interacting with the computer, but melding with it.

The connection between player and computer in a simulation game is a kind of *cybernetic circuit,* a continual feedback loop. Today the prefix "cyber-" has become so ubiquitous that its use has diffused to mean little more than "computer-related." But the word "cybernetics," from which the prefix was first taken, has a more distinct meaning. Norbert Wiener (1948) coined the term to describe a new general science of information processing and control. (He took it from the Greek word *kybernan,* meaning to steer or govern.) In particular, he was interested in studying, across disciplinary boundaries, processes of *feedback*: the ways in which systems—be they bodies, machines, or combinations of both—control and regulate themselves through circuits of information. As Steve J. Heims writes in his history, *The Cybernetics Group,*

> [The cybernetic] model replaced the traditional cause-and-effect relation of a stimulus leading to a response by a "circular causality" requiring negative feedback: A person reaches for a glass of water to pick it up, and as she extends her arm and hand is continuously informed (negative feedback)—by visual or proprioceptive sensations—how close the hand is to the glass and then guides the action accordingly, so as to achieve the goal of smoothly grabbing the glass. The process is circular because the position of the arm and hand achieved at one moment is part of the input information for the action at the next moment. If the circuit is intact, it regulates the process. To give another stock example, when a man is steering a ship, the person, the compass, the ship's engine, and the rudder are all part of the goal-directed system with feedback. The machine is part of the circuit. (Heims 1991, 15–16)

The constant interactivity in a simulation game—the perpetual feedback between a player's choice, the computer's almost instantaneous response, the player's response to that response, and so on—is a cybernetic loop, in which the line demarcating the end of the player's consciousness and the beginning of the computer's world blurs.

There are drawbacks to this merging of consciousness. When you're connected to the computer, it's easy to imagine you've transcended your physical body, to dismiss your flesh and blood as simply the "meat" your mind must inhabit, as the protagonist of *Neuromancer* puts it (Gibson 1984). This denial is a form of alienation, a refusal to recognize the material basis for your experience. The return of the repressed comes in the form of carpal tunnel syndrome, eyestrain, and other reminders that cyberspace remains rooted in physical existence.

But what the connection between player and computer enables is access to an otherwise unavailable perspective. In the collaboration between you and the computer, self and Other give way, forming what might be called a single *cyborg consciousness*. In her influential "Manifesto for Cyborgs" (1985), Donna Haraway proposed the figure of the cyborg—"a hybrid of machine and organism"—as an image that might help us make sense of the increasing interpenetration of technology and humanity under late capitalism. Haraway's point was that in this hypermechanized world we are all cyborgs. When you drive a car, the unit of driver-and-car becomes a kind of cyborg. When you turn on the TV, the connection of TV-to-viewer is a kind of cybernetic link. The man steering the ship in Heims's example is a cyborg. And most basically, since we all depend on technology to survive this postmodern world—to feed us, to shelter us, to comfort us—in a way we are all as much cyborgs as the Six Million Dollar Man.

Simulation games offer a singular opportunity to think through what it means to be a cyborg. Most of our engagements with technology are distracted, functional affairs—we drive a car to get somewhere; we watch TV to see what's on.[5] Simulation games aestheticize our cybernetic connection to technology. They turn it into a source of enjoyment and an object for contemplation. They give us a chance to luxuriate in the unfamiliar pleasures of rote computation and depersonalized perspective, and grasp the emotional contours of this worldview. To use the language of Clifford Geertz (1973) (borrowing from Flaubert), simulation games are a "sentimental education" in what it means to live among computers. Through the language of play, they teach you what it feels like to be a cyborg.[6]

Narrativizing Geography: Civilization II *as a "Spatial Story"*

So, what are the advantages to life as a cyborg? Why learn to think like a computer? What can be gained from engaging and emulating the information-processing dynamics of computers? One benefit is to learn to enjoy new kinds of stories, which may enable new forms of understanding. Unlike most of the stories we're used to hearing, a simulation doesn't have characters or a plot in the conventional sense. Instead, its primary narrative agent is *geography*. Simulation games tell a story few other media can: the drama of a map changing over time.

You begin *Civilization II* with a single band of prehistoric settlers, represented as a small figure with a shovel at the center of the main map, which takes up most of the computer screen. Terrain is delineated on this map by icons representing woods, rivers, plains, oceans, mountains, and so on. At the beginning of the game, however, almost all of the map is black; you don't get to learn what's out there until one of your units has explored the area. Gradually, as you expand your empire and send out scouting parties, the landscape is revealed. This process of exploration and discovery is one of the fundamental pleasures of *Civilization II*. It's what gives the game a sense of narrative momentum.

In their published dialogue "Nintendo and New World Travel Writing" (1995), Mary Fuller, an English Renaissance scholar, and Henry Jenkins, a cultural studies critic, compare two seemingly disparate genres that share a strikingly similar narrative structure. Nintendo games and New World travel narratives, like simulation games, are structured not by plot or character, but by the process of encountering, transforming, and mastering geography. Fuller writes, "[b]oth terms of our title evoke explorations and colonizations of space: the physical space navigated, mapped and mastered by European voyagers and travelers in the sixteenth and seventeenth centuries and the fictional, digitally projected space traversed, mapped, and mastered by players of Nintendo video games" (1995, 58). Borrowing from the work of Michel de Certeau (1984a, b), Jenkins labels these geographical narratives "spatial stories." He describes the process of geographic transformation as a transition from abstract "place" into concrete "space":

> For de Certeau (1984b), narrative involves the transformation of place into space (117–118). Places exist only in the abstract, as potential sites for narrative action, as locations that have not yet been colonized. Places constitute a "stability" which must be disrupted in order for stories to unfold. . . . Places become meaningful [within the story] only as they

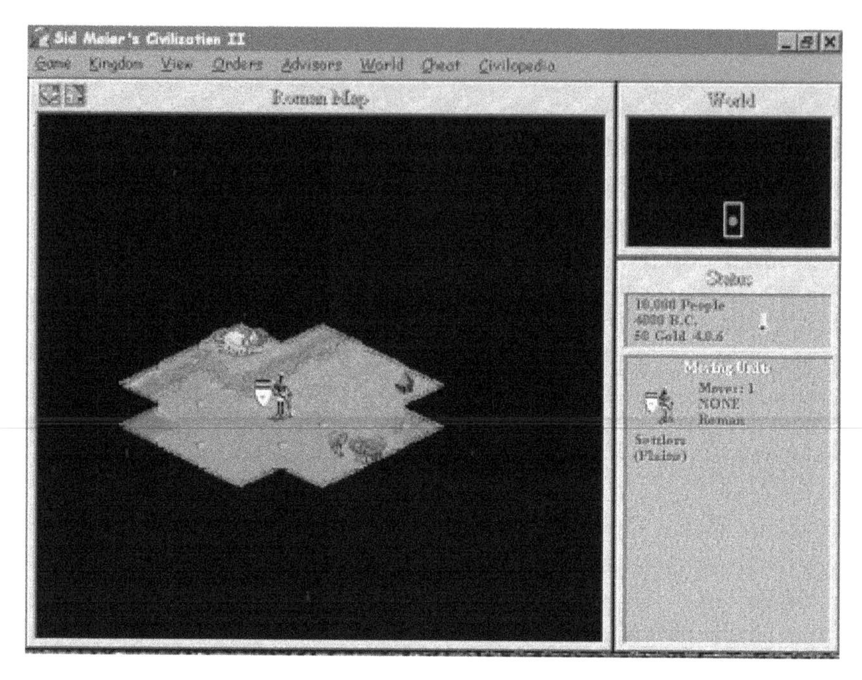

Fig. 6.1. You begin *Civilization II* with a single band of settlers, the unexplored terrain a black void.

> come into contact with narrative agents. . . . Spaces, on the other hand, are places that have been acted upon, explored, colonized. Spaces become the locations of narrative events. (Fuller and Jenkins 1995, 66)

Likewise, game play in *Civilization II* revolves around the continual transformation of place into space, as the blackness of the unknown gives way to specific terrain icons. As in New World narratives, the process of "colonization" is not simply a metaphor for cultural influence, but involves the literal establishment of new colonies by settlers (occasionally with the assistance of military force). Once cities are established, the surrounding land can be developed. By moving your settlers to the appropriate spot and choosing from the menu of "orders," you can build roads, irrigate farmland, drill mines, chop down trees, and eventually, as your civilization gains technology, build bridges and railroads. These transformations are graphically represented right on the map itself by progressively more elaborate icons. If you overdevelop, the map displays the consequences too: little

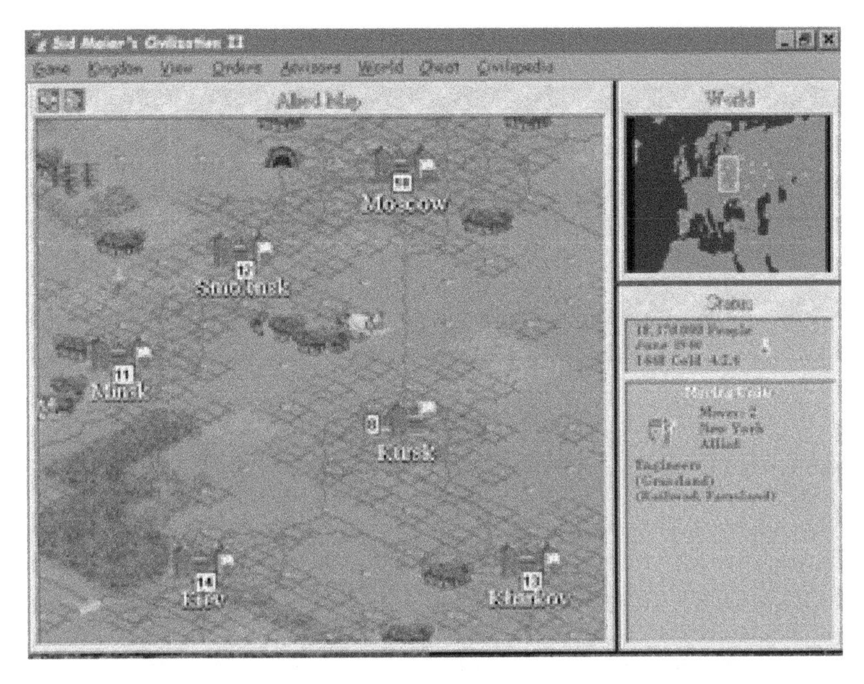

Fig. 6.2. As your civilization grows, you can transform the land with roads, bridges, mines, irrigation, and other "improvements."

death's-head icons appear on map squares, representing polluted areas that must be cleaned up.

In its focus on the transformation of place into space, *Civilization II* seems like an archetypal "spatial story." However, *Civilization II* differs from the geographic narratives Jenkins and Fuller describe in an important way, one that demonstrates the distinctive qualities of simulation games. In addition to the categories of space and place, Jenkins borrows two other terms from de Certeau, "maps" and "tours": "Maps are abstracted accounts of spatial relations ('the girl's room is next to the kitchen'), whereas tours are told from the point of view of the traveler/narrator ('You turn right and come into the living room') (De Certeau 1984b, 118–122). Maps document places; tours describe movements through spaces" (Fuller and Jenkins 1995, 66). Tours, in other words, are the subjective, personalized experiences of the spaces described abstractly in maps. You start your journey with a map. Then, as you navigate the geography, that abstract knowledge becomes the embodied firsthand experience of a tour. The maze of the Nintendo screen gives way to a familiar, continually retraced path that leads from the entrance to safety. The daunting expanse of the New World is structured by the personal account of one traveler's journey.

In the "spatial stories" Jenkins and Fuller discuss, then, the pleasure comes from two transitions, one involving geographic transformation, the other individual subjectivity. *Place* becomes *space* as unfamiliar geography is conquered through exploration and development. And *maps* become *tours* as abstract geography is subjectively situated in personal experience. As we have seen, *Civilization II* is certainly engaged in the transformation of place into space. But in simulation games the map never becomes a tour. The game screen documents the player's changes to the landscape, but these transformations are always represented in the abstract terms of the map. The point of view always remains an overhead, "God's-eye" perspective.

What's the import of this distinction? We might assume that the continued abstraction of the map would indicate a measure of detachment, compared to the ground-level engagement of a tour. But as already noted, simulation games seem singularly skilled at "sucking you in" to their peculiar kind of narrative. The difference is that the pleasure in simulation games comes from *experiencing space as a map*: at once claiming a place and retaining an abstracted sense of it. The spatial stories Fuller and Jenkins discuss respond to the challenge of narrativizing geography by "getting inside" the map—they zoom in from forest level so we can get to know the trees. Character may not be a primary criterion for these stories, but the stories still depend on individual subjective experience as the engine for their geographic narrative. Geography itself is not the protagonist; rather, the protagonist's experience of geography structures the narrative.

But simulation games tell an even more unusual story: the story of the map itself. Drawing a steady bead on the forest, they teach us how to follow, and enjoy, its transformations over time. We need never get distracted by the trees. Because simulation games fix the player in a depersonalized frame of mind, they can tell their story in the abstract, without ever bringing it to the level of individual experience. The map is not merely the environment for the story; it's the hero of the story.

The closest analogues I can think of to the distinct kind of spatial story that simulation games tell are works of "environmental history" such as William Cronon's *Changes in the Land* (1983). Cronon attempts to tell a version of American history from the perspective of the land, turning the earth itself into his protagonist. The limitations of the written word, however, make it difficult to fully treat an abstract entity as a character. You can't easily employ the devices normally used to engage the reader with a human protagonist. As a result, the book—like most works on geography—is still a rather dry read. It may offer a new perspective, but it can't engage the reader enough to give an emotional sense of what this perspective *feels like*.[7]

The clearest way to conceptualize space is not with words, but with images. A map captures the abstract contours of space; any verbal description begins the process of turning that map into a tour. This is why any good work of geography is full of maps; the reader is expected to continually check the words against the images, translating language back into visual understanding. Simulation games are a way to make the maps tell the whole story. As a still frame is to a movie, as a paragraph is to a novel, so is a map to a simulation game. Simulation games are maps-in-time, dramas that teach us how to think about structures of spatial relationships.[8]

Ideology

In one sense, every map is always already a tour. As geographer Denis Wood points out in *The Power of Maps* (1992), a map is the cumulative result of many subjective judgments. Maps always have a point of view. The ideological work of the "scientific," God's-eye view map is to make the traces of those subjective decisions disappear. Critics of computer games worry that the technological aura of computers further heightens this reification, leaving game players with the impression that they have encountered not just one version of the way the world works, but the one and only "objective" version (Brook and Boal 1995; Slouka 1995; Stoll 1995).

This perspective would leave little room to imagine resistance. But the structure of the computer gaming experience does allow for variant interpretations. You can win *Civilization II* in one of two ways. You can win by making war, wiping the other civilizations off the map and taking over the world. Or you can win through technological development, becoming the first civilization to colo-

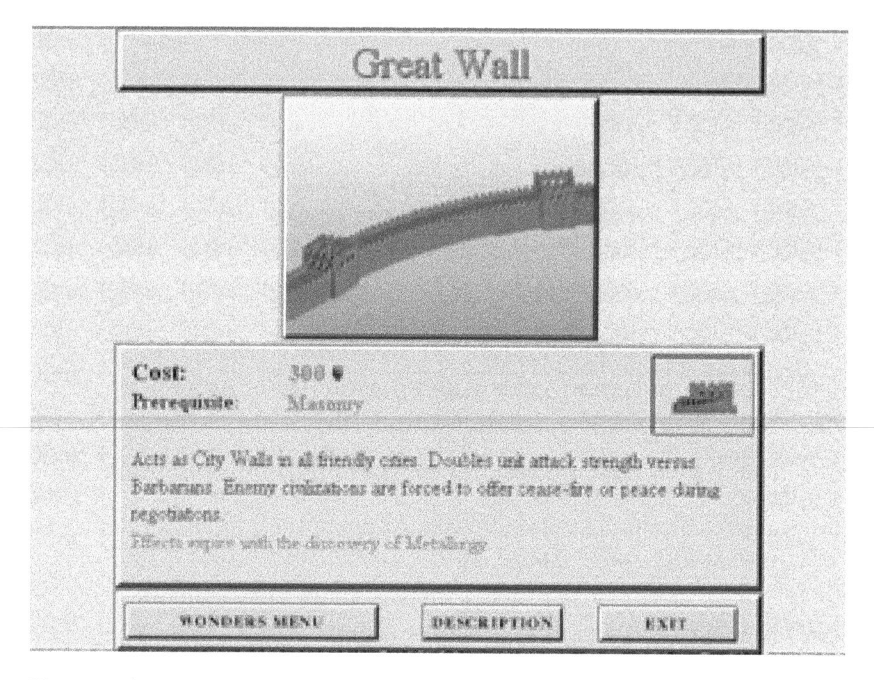

Fig. 6.3. The "Civilopedia" entry for the Great Wall.

nize another planet. I haven't emphasized the military aspect of *Civilization II* because I don't like war games all that much myself. They make me anxious. My strategy for winning *Civilization II* is to pour all my efforts into scientific research, so that my nation is the most technologically advanced. This allows me to be the first to build "wonders of the world," which, under the game's rules, force opponents to stay at peace with me. In the ancient world, the Great Wall of China does the trick; by modern times, I have to upgrade to the United Nations.

That's just one strategy for winning. I think it's probably the most effective—I get really high scores—but, judging from online discussions, it doesn't appear to be the most popular. Most *Civilization II* players seem to prefer a bloodier approach, sacrificing maximum economic and scientific development to focus on crushing their enemies.

The fact that more than one strategy will work—that there's no one "right" way to win the game—demonstrates the impressive flexibility of *Civilization II*. But there still remain baseline ideological assumptions that determine which strategies will win and which will lose.

And underlying the entire structure of the game, of course, is the notion that global coexistence is a matter of winning or losing.

There are disadvantages to never seeing the trees for the forest. *Civilization II*'s dynamic of depersonalization elides the violence of exploration, colonization, and development even more completely than the stories of individual conquest described by Fuller and Jenkins. Military units who fight and die in *Civilization II* disappear in a simple blip; native peoples who defend their homelands are inconveniences, "barbarian hordes" to be quickly disposed of.

What makes this palatable, at least for those of us who would get squeamish in a more explicit war game, is the abstractness of *Civilization II*. Any nation can be the colonizer, depending on who you pick to play. Barbarian hordes are never specific ethnicities; they're just generic natives. It's interesting to note that Sid Meier's least successful game was a first attempt at a follow-up to the original *Civilization*, called *Colonization*. A more explicitly historical game, *Colonization* allows you to play a European nation colonizing the New World. In addressing a more concrete and controversial historical subject, Meier is forced to complicate the Manifest Destiny ethos of *Civilization*. The Native American nations are differentiated, and behave in different ways. You can't win through simple genocide, but must trade and collaborate with at least some Native Americans some of the time. The result of this attempt at political sensitivity, however, is simply to highlight the violence and racism that are more successfully obscured in *Civilization*. There's no getting around the goal of *Colonization*: to colonize the New World. And while you have a choice of which European power to play, you can't choose to play a Native American nation.

Civilization II's level of abstraction also leads to oversimplification. The immense time span of *Civilization II* reifies historically specific, continually changing practices into transhistorical categories like "science," "religion," and "nation." Art and religion in *Civilization II* serve a purely functional role: to keep the people pacified. If you pursue faith and beauty at the expense of economic development, you're bound to get run over by less cultivated nations. Scientific research follows a path of rigid determinism; you always know in advance what you're going to discover next, and it pays to plan two or three inventions ahead. You can't play "the Jews" in *Civilization II*, or another diasporic people. The game assumes that "civilization" equals distinct political nation. There's no creolization in *Civilization II*, no hybridity, no forms of geopolitical organization before (or after) the rise of nationalism. Either you conquer your enemy, or your enemy conquers you. You can trade supplies and technology with your neighbors, but it's presumed that your national identities will remain distinct. Playing a single, unchanging entity from the Stone Age to space colonization turns the often slippery formation of nationhood into a kind of immutable racial destiny.

What to Do Once You've Conquered the World

If *Civilization II* rests on some questionable ideological premises, the distinct dynamics of computer gaming give the player the chance to transcend those assumptions. Computer games are designed to be played until they are mastered. You succeed by learning how the software is put together. Unlike a book or film that is engaged only once or twice, a computer game is played over and over until every subtlety is exposed, every hidden choice obvious to the savvy player. The moment the game loses its interest is when all its secrets have been discovered, its boundaries revealed. That's when the game can no longer suck you in. No game feels fresh forever; eventually you run up against the limits of its perspective, and move on to other games.[9] By learning from the limitations of *Civilization II* while exploiting the tools it offers, perhaps the next round of games can go further, challenging players' assumptions and expectations to create an even more compelling and rewarding interactive experience.

NOTES

1. I address the semiotics of *Sim City* in much greater detail in "Making Sense of Software: Computer Games and Interactive Textuality" (Friedman 1995).

2. The argument I will be making here is an extension of my discussion of subjectivity in "Making Sense of Software" (Friedman 1995). In that essay I describe the experience of playing *Sim City* as one of identifying with a *process*—with "the simulation itself" (85). In a simulation game, you don't imagine yourself as filling the shoes of a particular character on the screen, but rather, you see yourself as the entire screen, as the sum of all the forces

and influences that make up the world of the game. This essay extends that discussion by looking at how this perspective is, in a way, that of the computer itself.

3. I should clarify that in talking about "thinking like a computer," I don't mean to anthropomorphize, or to suggest that machines can "think" the way humans do. As artificial intelligence researchers have learned, often to their chagrin, computers can only systematically, methodically crunch numbers and follow algorithms, while human thinking is less linear, more fluid. My point is that using simulation games can help us intuitively grasp the very alien way computers process information, and so can help us recognize how our relationships with computers affect our own thoughts and feelings.

In describing computers as, in a sense, nonhuman actors with associated states of consciousness, I'm borrowing a technique of Bruno Latour, who in his novelistic history *Aramis, or The Love of Technology*, tells the story of a failed French experimental mass transit program from several perspectives—including that of the train itself. Latour writes,

> I have sought to show researchers in the social sciences that sociology is not the science of human beings alone—that it can welcome crowds of nonhumans with open arms, just as it welcomed the working masses in the nineteenth century. Our collective is woven together out of speaking subjects, perhaps, but subjects to which poor objects, our inferior brothers, are attached at all points. By opening up to include objects, the social bond would become less mysterious. (1996, viii)

Latour's conceit is one way to attempt to account for the interpenetration of our lives with technology, to make visible the often unnoticed role of technology in our daily experience and sense of selves.

4. Where, one may ask, in this confrontation between computer and player is the author of the software? In some sense, one could describe playing a computer game as learning to think like the *programmer*, rather than the computer. On the basic level of strategy, this may mean trying to divine Sid Meier's choices and prejudices, to figure out how he put the game together so as to play it more successfully. More generally, one could describe simulation games as an aestheticization of the programming process: a way to interact with and direct the computer, but at a remove. Many aspects of computer game play resemble the work of programming; the play-die-and-start-over rhythm of adventure games, for example, can be seen as a kind of debugging process. Programming, in fact, can often be as absorbing a task as gaming; both suck you into the logic of the computer. The programmer must also learn to "think like the computer" at a more technical level, structuring code in the rigid logic of binary circuits.

5. Actually one might argue that the pleasure many get out of driving for

its own sake and the enjoyment of watching TV no matter what's on (what Raymond Williams [1974] called "flow") are examples of similar aestheticizations of the cybernetic connection between person and machine. We might then say that just as these pleasures aestheticized previous cybernetic connections, simulation games do the same for our relationships with computers.

6. My reference here is to Clifford Geertz's famous essay, "Deep Play: Notes on the Balinese Cockfight" (1973), which discusses how a game can encapsulate and objectify a society's sense of lived social relations:

> Like any art form—for that, finally, is what we are dealing with—the cockfight renders ordinary, everyday experience comprehensible by presenting it in terms of acts and objects which have had their practical consequences removed and been reduced (or, if you prefer, raised) to the level of sheer appearances, where their meaning can be more powerfully articulated and more exactly perceived. (443)

This dynamic is particularly powerful because it is not just an intellectual exercise, but a visceral experience:

> What the cockfight says it says in a vocabulary of sentiment—the thrill of risk, the despair of loss, the pleasure of triumph. . . . Attending cockfights and participating in them is, for the Balinese, a kind of sentimental education. What he learns there is what his culture's ethos and his private sensibility (or, anyway, certain aspects of them) look like when spelled out externally in a collective text. (449)

7. One alternative might be to go ahead and treat an abstract object like a real protagonist, complete with an interior monologue. This is what Bruno Latour does in *Aramis*, as discussed above. But when one is discussing a subject as abstract as geography, even this move would likely remain a compromise with an inhospitable medium. In giving voice to geography, one risks anthropomorphization, falling back into the synecdochical trap of substituting the king for the land.

8. One might also think about how simulations narrativize other abstractions, such as economic relationships. In addition to being maps-in-time, simulations are also charts-in-time. One follows not only the central map in *Civilization II*, but also the various charts, graphs, and status screens that document the current state of each city's trade balance, food supply, productivity, and scientific research. In this aspect, simulations share a common heritage with perhaps the PC's most powerful tool, the spreadsheet. What the spreadsheet allows is precisely for a static object—in this case a chart—to become a dynamic demonstration of interconnections. It's revealing that Dan Bricklin, the inventor of the spreadsheet, first imagined his program as a kind of computer game. Computer industry historian Robert X. Cringely writes,

> What Bricklin really wanted was . . . a kind of very advanced calculator with a heads-up display similar to the weapons system controls on an F-14 fighter. Like Luke Skywalker jumping into the turret of the Millennium Falcon, Bricklin saw himself blasting out financials, locking onto profit and loss numbers that would appear suspended in space before him. It was to be a business tool cum video game, a Saturday Night Special for M.B.A.s, only the hardware technology didn't exist in those days to make it happen. (1992, 65)

So, of course, Bricklin used the metaphor of a sheet of rows and columns instead of a fighter cockpit. Simulation games, in a way, bring the user's interaction with data closer to Bricklin's original ideal.

9. I make a similar argument in "Making Sense of Software" (Friedman 1995).

WORKS CITED

The Best Ten Games of All Time. 1997. *PC Gamer*, May, 90–96.

Birkerts, Sven. 1994. *The Gutenberg Elegies: The Fate of Reading in an Electronic Age*. New York: Fawcett Columbine.

Brockman, John. 1996. *Digerati: Encounters with the Cyber Elite*. San Francisco: HardWired.

Brook, James, and Iain A. Boal. 1995. *Resisting the Virtual Life: The Culture and Politics of Information*. San Francisco: City Lights.

Cringely, Robert X. 1992. *Accidental Empires: How the Boys of Silicon Valley Make Their Millions, Battle Foreign Competition, and Still Can't Get a Date*. New York: Addison-Wesley.

Cronon, William. 1983. *Changes in the Land*. New York: Hill and Wang.

de Certeau, Michel. 1984a. *Heterologies: Discourse on the Other*. Trans. Brian Massumi. Minneapolis: University of Minnesota Press.

———. 1984b. *The Practice of Everyday Life*. Berkeley: University of California Press.

The Fifteen Most Significant Games of All Time. 1997. *PC Gamer*, May, 95.

Friedman, Ted. 1995. Making Sense of Software: Computer Games and Interactive Textuality. In *CyberSociety: Computer-Mediated-Communication and Community*, ed. Steven G. Jones. Thousand Oaks, CA: Sage.

Fuller, Mary, and Henry Jenkins. 1995. Nintendo and New World Travel Writing: A Dialogue. In *CyberSociety: Computer-Mediated-Communication and Community*, ed. Steven G. Jones. Thousand Oaks, CA: Sage.

Geertz, Clifford. 1973. Deep Play: Notes on the Balinese Cockfight. In *The Interpretation of Cultures*. New York: Basic Books.

Gibson, William. 1984. *Neuromancer*. New York: Ace Books.

Gilder, George. 1990. *Life after Television: The Coming Transformation of Media and American Life*. Whittle Direct Books.

Haraway, Donna. 1985. Manifesto for Cyborgs: Science, Technology and Socialist Feminism in the 1980s. *Socialist Review* 80:65–108.

Heims, Steve J. 1991. *The Cybernetics Group*. Cambridge: MIT Press.

Latour, Bruno. 1996. *Aramis, or The Love of Technology*. Trans. Catherine Porter. Cambridge: Harvard University Press.

Negroponte, Nicholas. 1995. *Being Digital*. New York: Vintage.

The One Hundred Fifty Best Games of All Time. 1996. *Computer Gaming World*, November, 64–80.

Shelley, Bruce. 1991. Manual for *Sid Meier's Civilization*. Hunt Valley, MD: MicroProse Software.

Slouka, Mark. 1995. *War of the Worlds: Cyberspace and the High-Tech Assault on Reality*. New York: Basic Books.

Stoll, Clifford. 1995. *Silicon Snake Oil: Second Thoughts on the Information Superhighway*. New York: Doubleday.

Wiener, Norbert. 1948. *Cybernetics: Or, Control and Communication in the Animal and the Machine*. New York: Wiley.

Williams, Raymond. 1974. *Television: Technology and Cultural Form*. Hanover, NH: Wesleyan University Press.

Wood, Denis. 1992. *The Power of Maps*. New York: Guilford.

Chapter Seven

Museum (Dis)Play
Imagining the Museum on CD-ROM

Alison Trope

> For a "Museum without Walls" is coming into being, and (now that the plastic arts have invented their own printing press) it will carry infinitely farther that revelation of the world of art, limited perforce, which the "real" museums offer us within their walls.
>
> —André Malraux

When André Malraux conceived the *musée imaginaire* (also known as the museum without walls) in 1947, he sought to open up traditional art history discourse and alter the institutional assumptions and approach practiced by conventional museums. Echoing the sentiments of Walter Benjamin's 1935 essay "The Work of Art in the Age of Mechanical Reproduction," Malraux argued that the technical conditions for color print reproductions in books and art catalogs necessarily reconfigured the museum's status and constitution, bringing a broad-based collection or repository of works to a wider audience. The imaginary museum would thereby construct an imaginary community in which works of art were autonomous and therefore no longer bound by architectural, institutional, or spatial constraints. Malraux's all-inclusive conception was, as Rosalind Krauss argues, "another way of writing 'modernism' " (1996, 344).[1] In a more contemporary, *post*modern context, however, Malraux's "museum without walls" has expanded and degenerated from a specialized, elitist art publication into a consumer marketplace flooded with postcards, posters, calendars, and T-shirts. The advent of digital storage (CD-ROMs) represents a continued transformation and evolution in the technological reproduction of artworks and, with it, a new conception of the museum without walls. Freed from the burden of re-creating an actual museum, the CD-ROM museum can play with the concept of the museum on the level of its space, architecture, design, holdings, and purpose. The simulation of the museum and its collection on CD-ROM therefore opens up possibilities for reconfiguring the institution's site-specific context and its inherent power relations.

The reconfiguration of exhibition, display, and reception, central to developments in CD-ROMs and other new media, also significantly parallels the revisionist politics of contemporary museological discourses. Beyond Malraux's museum without walls, I want to address an earlier critique of the museum: Le Corbusier's 1925 treatise, *L'art décoratif d'aujourd'hui*. In this polemical work Le Corbusier makes a clear distinction between good and bad museums:

> the true museum is one that contains everything, one able to give the whole picture of a past age. Such a museum would be truly dependable and honest; its value would lie in the choice that it offered, whether to accept or reject; it would allow one to understand the reasons why things were as they were, and would be a stimulant to improve them. Such a museum does not exist. (1987, 13)

With its capacity to store great quantities of data, construct a process of discovery, and permit a wide range of choice through its random access and relatively minimal outlay of capital, the CD-ROM museum approaches Le Corbusier's and Malraux's ideal. However, the notion of a "true museum," in Le Corbusier's words, or a "museum without walls," in Malraux's words, remains mired by the constraints of those who construct it and thereby control the dispersal of knowledge, namely, its curators and administrators. The museum has traditionally functioned as an arbiter of taste. The curator's, and implicitly the institution's, selection process and display strategies reflect what the cultural theorist Tony Bennett has called a "practice of showing and telling: that is, of exhibiting artifacts and/or persons in a manner calculated to embody and communicate specific cultural meanings and values" (1995, 6). In a traditional museum context the visitor is presumed to concur, or at least comply, with the curator's (and thereby the institution's) acquisition policies, display practices, point of view, and above all, taste.[2]

In the case of the CD-ROM museum, the curator-institution relationship is paralleled by the creative and business interactions and transactions of interface/conceptual designers, producers, and corporate distribution networks. The significance of this comparison lies in the (re)distribution of power in a potentially transformative sociocultural context. In both the museum and CD-ROM medium, the narrative and, to a large degree, the visitor/user's reception are constructed and manipulated from above. While the walls of the museum institution may have further crumbled with the advent of new media platforms and the erasure of site specificity, the power and economic relations persist even as they continually shift. The fantasy vision of a museum without walls, therefore, may be implicated in a potentially more deceptive museum with *invisible* walls (and invisible institutional power dynamics). Private industry (notably Bill Gates's Corbis Corporation) in many cases has begun to appropriate the function of collecting and exhibiting artwork to the public.[3] The new media industry and the Internet's corporate promoters and sponsors proffer a utopian vision of digital culture's democratic access to art and culture. Moreover, they construct a fantasy around the CD-ROM's intrinsic freedom of exploration, privileging the medium's random access and its use-value in the private, domestic sphere.

Despite such utopian fantasies of the public's consumer access and the user's random access to information, many have argued that the promise of interactivity has been overrated, and should be mediated by a healthy degree of skepticism. The media researcher, educator, and curator Erkki Huhtamo claims that the freedom of new technology can be inherently deceptive: "the potential that interactivity offers—a personalized, active, and intimate relationship with media, in an associative coupling of different sensory registers, multiform retrieval of information—has been harnessed into the service of ideologically guaranteed and commercially calculated formulas" (1996, 309).[4] Huhtamo persuasively frames his argument about CD-ROM technology vis-à-vis core issues of poststructuralist theory by privileging an open and personal interaction with the technology over a unified formulaic presentation of material.

Michael Nash, former president and creative director of Inscape, similarly calls for and privileges such open-ended access. In his article

"Vision after Television: Technocultural Convergence, Hypermedia, and the New Media Arts Field," Nash claims that "hypermedia offers a potential quantum leap in the dialogue between mind and inscape" (1996, 392).[5] For Nash, the "inscape" represents a compelling inner landscape uncovered through a provocative, metaphoric, and immersive experience with multimedia. Huhtamo foregrounds the potential of children's software (or childware) to provide this kind of access or inscape by working against formula and significantly experimenting on the level of form and content, while Nash foregrounds creative game titles such as the Residents' *Freak Show* for their exploration of backstage and behind-the-scenes spaces. These examples reflect an often unrealized potential of multimedia to encourage the user's participation on a mental as well as a more transparent mechanical level. In fostering this dual and active mode of participation, the CD-ROM can ideally provide a context for accessing what is otherwise generally elided—the experiential and idiosyncratic nature of the multimedia encounter.

Most CDs, however, continue to be produced according to familiar and commercially driven formulas. The majority of formulaic CD game and reference titles are framed by parameters and generic conventions; similarly, the typical CD-ROM museum upholds convention by promoting an institutional and curatorial narrative and thereby structuring a visitor or user's experience. The limited number of CDs on the retail market that represent the museum and/or the museum collection tend to feature and privilege a repository or catalog of artifacts rather than exploring the museum as a cultural entity, institution, architectural space, or historical public sphere. On the surface, the most popular format replicates a text-based catalog or "coffee table" art book that features textual analysis alongside photographic reproductions (including full-size images as well as details) of artworks. Even fewer CD-ROM museums approach the potential for immersion and play outlined by Huhtamo and Nash to explore backstage and behind-the-scenes spaces or to insert one's own subjectivity into the institutional context. In this analysis, however, I am not so much interested in judging one platform against another as in understanding how and why the museum and the museum collection are adapted and represented in these varying frameworks, and further, how the reality of the exhibition venue or site becomes reframed and imagined as a conceptual idea or metaphor. The tension between a formulaic and nonformulaic or alternative mode of CD-ROM production has particularly intriguing implications for the CD-ROM museum, whose representational strategies can reveal the way the medium interacts with structures of power. These strategies, in turn, reveal conventions that will undoubtedly influence new media and their dialogue with the museum as well as the contemporary art world and art market.

In its ideal form, the CD-ROM museum offers a virtual visit with exclusive access to a museum's collection. The advantage of such access is evident. The CD-ROM museum, through spatial and temporal models, expands the print or catalog reproduction and reconstruction of the world's art collections. While only a fraction of a museum's collection can be displayed at any one time within a museum's walls, a CD-ROM's storage and reference capacity far exceeds such architectural limitations. Fragile, light-sensitive, and deteriorating works previously unavailable for public viewing (including prints, drawings, and in some cases tapestries and other textiles) are now available for appreciation, examination, and comparative study through CD-ROM technology.[6] In addition, artworks housed in other states or countries (especially those that rarely travel, such as the Barnes Collection) are also made available without extensive travel or expense. The CD-ROM museum, therefore, upholds an enduring travelogue tradition, in some sense, playing a similar role to the Lumière brothers' early actualities and travel films. They bring the world, or in this case, the world's museums and masterpieces, both spatially and temporally into the spectator/user's world, giving him/her a sense of possessing the object and "being there."[7] The CD format, especially in contrast to the average overcrowded blockbuster exhibit, may also provide a more pure, or at least more satisfactory, level of contemplation—a personal re-encounter with the work of art. In some of these cases, then, the copy may arguably equal (if not be preferable to) the experience of the original. However, according to CD-ROM museum producers and promoters, these titles are not trying "to be," duplicate, or replace the real museum experience. Rather, they are meant to serve as reference, supplement, inspiration, and (*only* in rare cases) surrogate for the real visit.[8] Given this scenario, what exactly *is* the cultural significance of the CD-ROM medium in relation to the museum? How does a site-specific institution function in a non–site-specific context? And what will be the impact of the CD-ROM's existence, use, and conceptual design on future relations between multimedia technologies and the museum as a cultural institution?

Playing It Safe: Convention Rules the Retail CD-ROM

Since CD-ROMs entered the new media retail market, a select group of art institutions, often in collaboration with software developers and/or major corporations, have utilized the medium to create a storage and display context for their artifacts. In 1993 Microsoft produced *Microsoft Art Gallery: The Collection of the National Gallery, London*. That same year, another Northern California software developer, Digital Collections, produced its first Mac-only version of New York's Frick Collection. The company, since incorporated as Digital Arts and Sciences Corporation, subsequently produced several other fine art CD-ROMs through 1996 that feature selected paintings from museums, foundations, and private collections, including the Brooklyn Art Museum's Ancient Egyptian Collection, Harvard University's Fogg Art Museum, the State Russian Museum in St. Petersburg, the Joe Price Collection of Japanese painting, and the Robert Mapplethorpe Foundation Collection. In 1995 Bill Gates's Corbis Corporation produced *A Passion for Art: Renoir, Cézanne, Matisse and Dr. Barnes* in collaboration with the Barnes Foundation. That same year, the Louvre was reconstructed on CD-ROM (*Le Louvre: The Palace and Its Paintings*) through a joint venture of Réunion des Musées Nationaux, Montparnasse Multimedia, and BMG Interactive. In 1996 the Smithsonian Institution put its National Museum of American Art on a CD-ROM distributed by Macmillan Digital USA. The catalog CD-ROM museums (mentioned above), like most printed catalogs or art books, feature a single museum or art collection. Other fine art CD titles, also comparable to printed books, feature the work of a single master artist (Cézanne, Botticelli, da Vinci), or a single historical site (the Sistine Chapel ceiling, the Vatican, the Hermitage). These titles are generally marketed as educational and reference resources.

The catalog format is the most popular among museum CD titles. It provides the cheapest, easiest, and most cost-effective means of communicating information about artworks, artists, and movements (historical, national, stylistic, etc.). *Microsoft Art Gallery* (1993), one of the first corporate-backed museum CD titles, can be credited with establishing some of the initial conventions of the CD-ROM museum. Like other Microsoft ("Microsoft Home") reference and encyclopedic titles (e.g., *Encarta, Cinemania*), *Art Gallery* follows a fairly elementary and straightforward pedagogical agenda. The opening menu provides a brief history of the National Gallery as a site (when it was founded, where it is located, what it holds, etc.) in addition to listing five specified areas open to further exploration: "Artists' Lives," "Historical Atlas," "Picture Types," "General Reference," and "Guided Tours." Subsequent museum titles have tended to follow a similar pedagogical agenda with only slight variations.

This pedagogical approach, the use of the CD-ROM as encyclopedic reference and/or teaching aid, historically conforms to the original goals of the museum institution. Tony Bennett, among others, positions the birth of the museum in relation to eighteenth- and nineteenth-century reform movements as well as the development of other social and cultural institutions designed to curb and construct accepted patterns of behavior. Along these lines, Pierre Bourdieu has argued that the fine art museum contributes toward the maintenance of the social order in constructing a sense of distinction between those who possess and those who lack cultural capital (Bourdieu and Darbel, 1990). The educational function of the museum, however, has been complicated throughout its history by a need to attract visitors and compete with other leisure activities. Most museum historians and theorists agree that, to a large extent, the educational function of the museum space and its artifacts was supplanted by its entertainment function early on.[9] Given its status as a consumer retail product, the typical CD-ROM reference title confronts a similar tension between its educational and entertainment function and goals. The role of the CD-ROM museum as reformer, arbiter, and constructor of cultural capital therefore not only parallels the museum's initial intentions, but arguably reaches a new (though not necessarily wider) public by entering the private sphere of the home and the personal computer.

In order to enhance the CD-ROM museum's pedagogical agenda and go beyond an encyclopedic catalog of works, many of these CD titles also attempt to conjure the museum environment through conventions that deploy spatial and temporal metaphors. These metaphors (the gallery, the library or archive, the museum tour, the slide show, etc.) not only represent a list of options on the interface; they

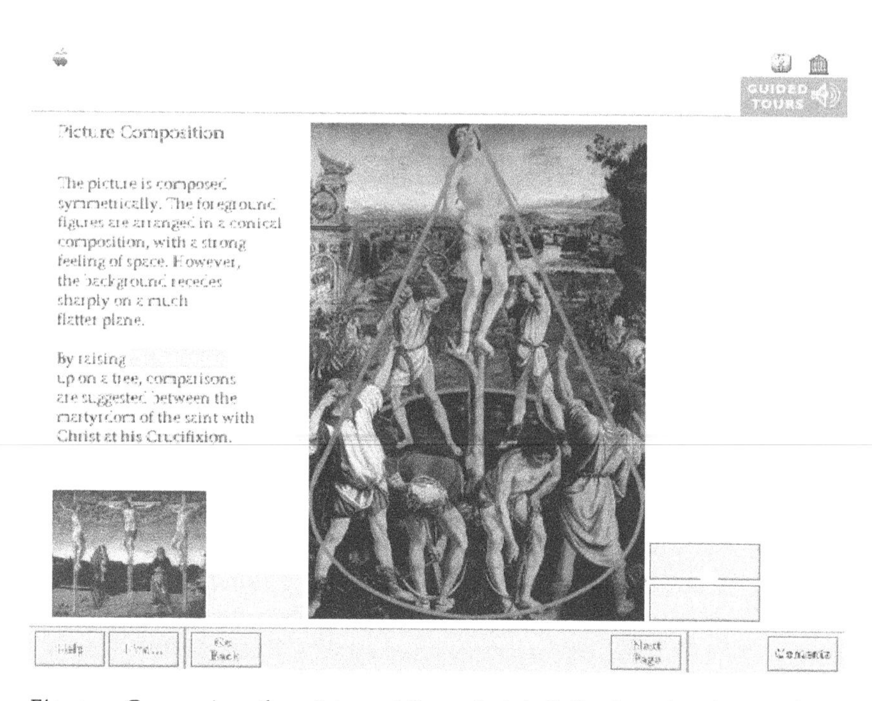

Fig. 7.1. Composing the picture: *Microsoft Art Gallery*'s animation explains perspective.

evoke a familiar narrative as well as a familiar space within the larger space of the museum's whole. From 1993's *Microsoft Art Gallery* to more recent titles such as the Smithsonian's *National Museum of American Art*, the opening screen options serve as the museum "entrance" or narrative introduction—complete with a welcome as well as a map or breakdown of the museum's (and the CD-ROM's) contents. The Smithsonian title replicates a classical museum facade by using Doric columns as an "entrance" and overall interface design. In the majority of titles, the following screen in the hyperlink chain generally features a more in-depth exploration of a particular narrative subcategory listed on the opening screen (e.g., "Timeline," "Historical Atlas," or "Maps"; "Paintings" or "The Collections"; "Schools" or "Picture Types"; "Tours"; "Archive," "Library," or "Index," etc.).

The labels for these various objects and spatial metaphors may differ slightly from title to title, but the underlying approaches similarly foreground a temporal-historical trajectory. The guided tour, for example, implicitly suggests an imaginary journey through the museum galleries. The audio portion of these tours, generally "performed" in an authoritative curatorial voice-over, further corresponds to the tape-recorded audio tours that have become a *de rigueur* component of most blockbuster museum exhibits. The typical CD-ROM museum tour recounts the institution's history, reviews its collections, and highlights particular artists, techniques, and recurring subjects and genres. Such tours clearly manifest an appropriation of the real museum's pedagogical props and narrative storytelling strategies. The metaphors creatively suggest time travel, geographical journeys and tours through familiar spaces and social institutions, thereby offering an entertaining journey through, or at least a distraction from, the seemingly boundless information stored on the CD-ROM. As a storage site, then, the CD-ROM relies on the familiarity of the tour and other metaphors suggestive of movement to offer a recognizable and interesting site for the works of art it "exhibits."

In order to construct an exhibition context, the principal CD-ROM museums (the National Gallery, the National Museum of American Art, the Barnes Collection, and to a lesser extent, the Louvre) feature animation sequences and video-motion simulations (generally including audio).[10] Movement through space and time is achieved through established formal codes of film, video, and animation. In the gallery tours, the user is generally encouraged to sit back and watch as the digitized images of paintings and galleries pass before him/her. The animated sequences simulate familiar camera movements, including pans, tilts, zooms, and close-ups. Other animations use superimposition to create a collage effect, thereby juxtaposing works by the same artist or works with the same stylistic or thematic tone.

Most programs also construct slide shows, where images pass in front of the screen individually, and the user can typically control the browsing speed. Animations are also used to construct grids over original works to expose and thereby explain the painting's underlying composition or perspective. Shading or highlighting segments of paintings is another technique used to draw attention to an imaginary line between two halves of a given work, thereby revealing a particular painting process, the evolution a particular work has undergone, the restoration of a key painting, and so forth. In most of these animated scenarios, however, the user is only moderately active, generally only instigating the start or stop of an animation process. The catalog portion of the CD-ROM museum, therefore, potentially provides more room for interaction than the tour component. While both catalog and tour are firmly constructed by the authority and knowledge of the curator, the catalog allows the user/visitor to encounter the works through his/her own choice and at his/her own pace.

In privileging the catalog format (generally featuring preconstituted tours), most of the museum collection titles currently available tend to uphold and recycle a conventional narrative and metaphor of the museum without pushing limits or engaging current museum theory or discourse. They fail, in other words, to explore the institution and its arenas as constructs, or further, to engage the museum and its discourses in metaphoric play. In part this can be attributed to the institutional priorities and expense involved in the production of a CD-ROM title. Besides economic factors and technical constraints, the formulaic approach that omits the museum context may also be interpreted as a conscious decision on the part of designers and museum institutions alike to avoid direct comparisons between the virtual visit and the experience of a real museum visit.[11] The traditional emphasis on the repository or catalog of artworks can thereby be read in the context of what the institution, in many cases, conservatively values—the autonomous object over its site specificity.

At the risk of oversimplifying, many fine art museums still prioritize their collection, preservation, and conservation activities over their educational activities. Since CD-ROM technology tends to be classified in the latter category, production of CD-ROM museum titles has not been actively pursued by the majority of museum institutions.[12] At the same time, however, pressures on museums that particularly rely on public funding have necessitated a reevaluation of museum services and exhibition politics. The museum is therefore caught in a double bind. On the one hand, it seeks to fulfill its traditional mission by serving as public guardian for the arts, while on the other, it must respond to social and cultural changes that have produced a new audience with increasingly high-tech expectations. The members of this new audience also demonstrate a greater interest in behind-the-scenes operations and non–collections-based displays and facilities (Macdonald, 1992). Although pressure to employ CD-ROMs and other new technologies to engage this new audience may be pervasive, an institution's final decision to develop and produce a CD-ROM remains motivated primarily by economic factors.

While nearly every museum in the world can now claim its own

Web site, relatively few institutions have put their collections on CD-ROM.[13] Budget, lack of funding, and problems with sales and distribution for such a venture are obvious factors inhibiting production. The public and trade press perception of art-oriented CDs has been another factor. Many of these CDs are marginalized in retail venues and have been labeled "fringeware" by the multimedia trade press (Huhtamo, 1996). The expectation that major book store chains such as Borders would act as a retail outlet for fine art CD titles has recently vanished, and in the last year, nearly every New York publishing house has closed its CD-ROM division. Changes in technology and developments on the Web have continued to decimate the distribution potential for the majority of consumer reference CD-ROMs.

In the past four years the production costs of most CD-ROM museums in the retail market have ranged from approximately $500,000 to as much as $1.5 million. Typically, a museum title will sell in a fairly limited number of retail stores and range in price from thirty to fifty dollars.[14] The retail sales of such titles have not been particularly encouraging for future development in nonprofit art institutions currently struggling for funding in both the private and public sector. Private companies like Corbis, meanwhile, claim a higher success rate. *Microsoft Art Gallery* and *A Passion for Art* (the Barnes Collection) have sold well for the Corbis Corporation. Other products (produced by companies besides Corbis), however, have had more difficulty recouping their initial investment in production and labor.[15]

Aside from the retail CD-ROM, a popular alternative currently pursued by many museums is the in-house digital lab or resource center where visitors can explore the arts as well as their historical, social, and cultural influences through a variety of media ranging from touch-screen computer kiosks (housing CD-ROMs or other high-capacity digital storage) to Web page access to video installations.[16] In the future, museums are also likely to produce in-house titles that can accompany specific exhibits, thereby functioning as merchandise—a souvenir that invokes a particular experience—rather than an attempt to map an entire institution's holdings. For example, the Metropolitan Museum of Art has had great success with its 1996 title accompanying the *Splendors of Imperial China* exhibit, as did the Smithsonian's CD-ROM accompanying its recent *150 Years of America's Smithsonian* traveling exhibition. The *Leonardo da Vinci Codex* exhibition in the fall of 1996 at New York's Museum of Natural History further confirmed the success of the CD-ROM as both substitute and supplement for the more traditional book catalog popularly purchased at such exhibits.

If the Corbis Corporation's archive of seventeen million images and successful CD-ROM sales are any indication, access to digitally reproduced artwork will continue to grow in the future via the Internet and other new media forms.[17] The rights to electronic images, however, will undoubtedly affect this access. Institutions have been selling the rights to their collections to private corporations such as Digital Collections and Corbis for the past few years. Bill Gates's acquisition of the electronic rights to various art and photography collections may serve as a model for the industry in the future. The question that predominates is whether Gates will offer these works for public display and free access in the tradition of the public museum, or will more predictably combine his philanthropic impulses with his profit motives. In the meantime, it remains uncertain how the technological appropriation of a museum's holdings will directly affect the museum as a cultural and social venue as well as publicly funded institution. Irrespective of legal rights and holdings, however, it seems safe to say that exhibition and access will continue to be dependent on and determined by the institution, whether it be corporate or nonprofit.[18] Both the profit and nonprofit scenarios will likely continue to follow tradition by employing an authoritative, curatorial voice that constructs fixed narratives and value systems.

Rather than pursuing or satisfying the fantasies and possibilities of an interactive museum without walls, the typical CD-ROM museum clearly reflects and replicates the traditional institution's cultural power structures. Continued mergers between (or takeovers of) museum institutions and the corporate world will predictably mimic this same power dynamic. However, poststructural shifts over time may continue to offer an alternative approach or way out of this institutional bind. Contemporary museological theory, for example, has attempted to rewrite traditional conceptions of the museum, reconfigure institutional power dynamics, and reframe the position and status of the museum visitor. After examining various economic, institutional, and curatorial barriers in the production of CD-ROM museums, I want to address the current theories of museological discourse to see how they might inform future prospects of multimedia's dialogue with the museum.

Bringing the Museum into Play: An Overview of the New Museology

Since the mid to late 1980s, the critical and theoretical inquiry into the museum has been substantially reframed. A new body of literature on museums has attempted to deconstruct traditional associations attached to the institution, turning away from a strict art historical approach that examines objects housed *within its* walls. Instead, this new museology focuses on the museum *itself* as the object of its analysis. Contemporary museology thereby echoes Le Corbusier's and Malraux's desire to self-consciously question the museum's identity, mission, rhetoric, and textual status. Revisionist writings on exhibition, representation, and display question whether the museum, as bastion of conservative institutionalism, can integrate alternative approaches that could potentially undermine or challenge its cultural authority. The interrogation of what constitutes cultural authority is, indeed, at the heart of museum studies.

Traditional museology, as a science of museums, surveys the histories and theories associated with the everyday practices of museum institutions from exhibition design, research, and management systems to collection, preservation, and education. Through the organization and design of exhibition space as well as its strategies of classification and display, the museum, since its inception, has employed narrative conventions in order to manage space and time as well as make the museum visit both educational and entertaining for the average visitor.[19] As a site built for reflection, contemplation, and edification, a museum strategically employs its exhibition in order to provide context, meaning, value, and often a mythic view of culture, history, and the nation. The architectural plans and layout of a museum's interior, the design of exhibitions, and the targeting and orchestration of audiences have become a science supported by statistics, sociological studies, and museum training programs worldwide (Edson, 1995).

Over the course of its two-hundred-year history, the museum has maintained a relatively fixed set of institutional conventions pertaining to exhibition and reception. Tony Bennett labels the museum a "backteller," meaning it has the prophetic power or hindsight to frame a visitor's experience in such a way that s/he witnesses a virtual past never before seen or accessible. According to Bennett, the power of backtelling—the bestowal of this "socially coded visibility"—comprises the "narrative machinery" of the museum. The allusion to machinery is meant to conjure an image of forward progress and evolution. The museum visit is therefore meant to unfold in a linear fashion, thereby promoting a performative function—a model of "organized walking through evolutionary time" (Bennett, 1995, 10). Pedagogical props and instructional aids such as wall texts, guidebooks, and audio recordings further serve the linear, tour-oriented flow by encouraging a progressive and timely movement through the space. Such props also serve the narrative function of the museum's displays and spatial organization by telling a linear story. By contrast, the new museology challenges conventional patterns of reception and power/knowledge relations that otherwise efface a visitor's presence and (inter)action in the museum.

Beyond this, the new museology questions the fixed body of knowledge purported by the museum institution. The typical museum narrative builds a fictive coherence for the spectator/visitor, leading him/her through a series of culturally and historically coded times and places. As Eugenio Donato has written in an essay on Flaubert's *Bouvard and Pécuchet,*

> The set of objects the Museum displays is sustained only by the fiction that they somehow constitute a coherent representational universe. The fiction is that a repeated metonymic displacement of fragment for totality, object to label, series of objects to series of labels, can still produce a representation which is somehow adequate to a nonlinguistic universe. Such a fiction is the result of an uncritical belief in the notion that ordering and classifying, that is to say, the spatial juxtaposition of fragments, can produce a representational understanding of the world. (1979, 223)

Clearly the museum's tendency toward narrative closure and a fictive coherence is open to criticism. To challenge the museum's conventions, the new museology deploys a poststructural critique, calling on the museum to unveil its institutional facade and foreground the significance of its space in relation to what is exhibited. The poststructural critique as applied to the museum engages both the institution as a social and cultural entity and its politics of address and reception. Traditionally the museum's politics of representation have upheld the values and visions of the dominant culture and ideology. Deconstructing the museum institution reveals how various ethnicities, classes, and genders have been positioned by display and exhibition. A more self-reflexive reception to spatial configuration and exhibition design can potentially foster a conceptual awareness of the museum institution and its holdings on a meaningful level. The visitor, ideally, is made aware of the constructedness of the museum, and is in turn able to actively participate in the process of constructing knowledge.

The question remains how to effectively and continually alter this reception. Attempts to apply these theories to the actual (or virtual) museum experience expose the frequent gulf between theory and practice. In order to effect a shift in the museum's unified representation and address the challenges posed by the new museology, some museums have consciously and self-reflexively revamped their exhibition spaces by foregrounding display practices, pedagogical tools, and public relations strategies. The value of such a self-reflexive approach lies in the new and reconfigured access it affords the museum visitor. Ideally the museum's behind-the-scenes mode of production is made transparent; the visitor is thereby empowered to de- and reconstruct the museum and its various discursive modalities. The visitor thereby becomes a player in the museum's process of construction. While many institutions have responded in one way or another to the concerns of the new museology, their responses have generally been relegated to individual exhibits or special retrospectives. Rarely has the new museology been applied on any kind of sustained basis. This is most likely due to the impractical expenditure of labor and production costs incurred in constant reconstruction. Beyond spatial reconstruction, however, an institution can more readily engage new museology by providing or at least fostering a more direct visitor interaction with the museum space as well as its holdings.

The critique of the museum may significantly reemerge in relation to the arts, the museum, and other cultural institutions as they face pressures to adapt to a burgeoning world of multimedia. In this context the CD-ROM museum offers a platform to rethink the challenges raised by museum studies. CD-ROMs can serve as a useful prototype to envision future interchanges between museums, multimedia, and their visitors/users. In its current state of production, however, the catalog CD museum format (often enhanced by animated tours) still has its drawbacks. The omission of context, for example, is problematic on several levels. First, the spatial context (the way an object is positioned on a wall, its juxtaposition and therefore relation to other objects, its framing, etc.) can potentially, if not definitively, affect an object's meaning. In addition, the institutional space of the museum itself holds meaning both in relation to the works it houses and independently as a social and cultural public sphere. While the majority of CD-ROM museums have elided these contextual issues, a few titles have attempted to take them into account by playing with space and the institution as construct. Such play, in turn, activates the possibilities engaged by the new museology, and further offers a sphere where museological theory and practice may significantly intersect.

The Museum as Play-Ground: Navigating the Institution

The more experimental museum titles, distributed and in most cases produced by the Corbis Corporation (*A Passion for Art: Renoir, Cézanne, Matisse and Dr. Barnes; Paul Cézanne: Portrait of My World;* and *Leonardo da Vinci*), go beyond a mere filmic use of animation to construct a virtual environment where the user/visitor can actually navigate the exhibition space. The movement through space and time is more than metaphoric on these CD titles; it is determined by the (inter)action of the user/visitor. The Barnes Collection served as the prototype for Corbis's other navigable interfaces (on the level of design as well as expense).[20] In the "Gallery" section of the CD-ROM, the user/visitor can explore selected masterpieces room by room as they are laid out in the actual Barnes building, located outside Philadelphia. The selections/paintings that the user/visitor can access in depth are highlighted in color, and further designated by the cursor's common transformation into the symbol of a hand. The other works remain in black and white.[21] In the lower margin of the screen, floor plans of the first and second floors indicate which room the user/visitor currently occupies and which wall s/he is facing. In the upper margin of each screen the room number is identified as well as the direction the user/visitor is pointing (north, south, east, west). In order to navigate through the space the user can either click on a numbered room of the floor plan for immediate access, or use the arrow cursor to move through the gallery. The fluidity of the user's movement on the Barnes title is restricted, to a large degree, by the state of the technology. Unlike the future prospects of real-time/3-D, which offer a more pan-

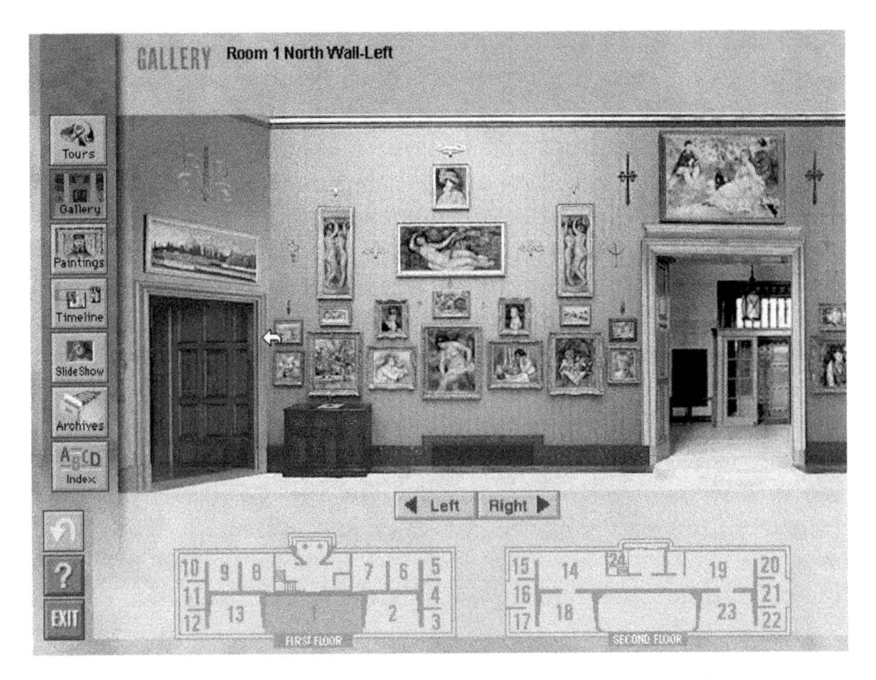

Fig. 7.2. Touring the space: the user/visitor navigates through the Barnes Collection by clicking from room to room.

oramic view and fluid movement through space, the movement on the Barnes title more accurately resembles a (filmic) cut from one perspective or ninety-degree angle (i.e., wall of paintings) to another. This less developed type of navigation distances the user/visitor from the real experience of moving through a gallery space.

Other Corbis products also foreground spatial navigation. The Cézanne title re-creates the artist's studio and features paintings as well as Cézanne's personal effects (paintbrushes, etc.). Like the Barnes title, this CD-ROM uses color as opposed to black and white to indicate the paintings and objects that act as hyperlinks to further information. The Leonardo da Vinci CD, one of Corbis's most recent products, constructs a faux rotunda and gallery space based on a museum da Vinci had designed but which was never realized. The eight gallery spaces accessible from the rotunda's three archways feature paintings, manuscripts, documents, lost works, and engineering plans as well as studies and drawings of nature and humanity. In all of these Corbis titles, the gallery navigation provides the user access to the paintings and other works in the collection. The gallery navigation therefore serves as a link between points of reference about individual works, artists' lives, historical time lines, and geographical maps. It also provides the concrete spatial context of the museum's walls, and thereby potentially offers more than an imaginary mental-scape for the artwork.

The navigation model therefore draws attention to (even privileges, in some cases) the institution as essential to the artwork's overall meaning and import. For the Barnes Collection, for example, the actual gallery context and exploration are particularly significant. Albert Barnes's theories of aesthetics and art appreciation were central to his exhibition practices. His concept of "wall ensembles," Barnes's catchall exhibition approach based on the relation of single works to the larger ensemble, dictated his display strategies. While the CD-ROM provides a more accurate representation and re-creation of these wall ensembles in comparison to the print-text renditions of the collection, the obvious question remains whether the spatial navigation component adds to the user's conceptual understanding and experience of the actual site, and whether the experience of the actual site is, in the end, an important one.

Like the spatial metaphors and animations of these CD titles, navigating a constructed space provides yet another means of creating an entertaining journey through the stored material on the CD. On one level, spatial navigation implies an interaction between the user and the material, thereby potentially engaging the specificity and potential for choice offered and hyped by the medium.[22] On another level, the degree of interaction offered by spatial navigation may be read by future generations as a mere novelty. The popularity of multimedia, and the CD-ROM in particular, may be derivative of a sense of wonder and attraction attached to the technology itself. The "bells and whistles" of this new technology offer the user a new, or at least redefined, experience. Rather than simply reading an encyclopedic description or definition, the CD-ROM user clicks on images, buttons, and text to effect sound and movement.

The wonder attached to this experience with technology may be analogous to the concept of the "cinema of attraction(s)," coined by the film historian Tom Gunning to describe the novelty period of prenarrative film. Gunning (1986) argues that the power of early cinema resided in its ability to show, to technologically harness visibility, to be an exhibitionist medium rather than focus on storytelling and the fictional diegesis.[23] CD-ROM technology and user-interface design arguably function along similar lines, by reading the technology as an attraction and story as mere pretext or diegetic backdrop. The emphasis on exhibition display (i.e., attraction) and the novelty of new technology, however, should not be discounted as a primitive phase in a larger evolutionary model of multimedia or CD-ROM development. Instead we must examine a wide range of models and approaches in order to reveal the complex, dynamic, and dialectical development of the medium in general, as well as the CD-ROM museum more specifically.[24]

Playing the Field: Behind the Scenes of the Museum

Certain CD museum titles have successfully and experimentally harnessed the medium's novelty and attraction potential by attempting to construct alternative narratives, and thereby providing access to a different conception and understanding of the museum. In permitting the user/visitor relatively open access and free rein, the navigational models employed by Corbis seemingly allow the visitor to construct his/her own museum tour. However, while the navigation platform explores the spatial context of the museum to a certain degree, its re-creation of the visitor's actual motion through time and space is, at present, less than fluid. Technological limitations and high-end production and labor costs hinder the provocative potential of the user-navigation platform, as do conventional and formulaic approaches to the user and his/her temporal and spatial movement.[25]

The interactive component, however, is further enhanced on those titles that encourage the user/visitor to make a portfolio (*National Museum of American Art*) or a slide show (*A Passion for Art*) of selected works, or in other cases, encourage the user/visitor to apply the provided resources as a reference tool.[26] In general, the Smithsonian's *National Museum of American Art*'s "Gallery" section (composed of the "Media Center," "Director's Choice," and the "Browser") provides the most innovative and self-reflexive approach to the representation of the museum. The "Media Center" supplies an overview (through a choice of the following media: video, audio, album, slides) of the artists featured in the museum, their work processes, techniques, and personal opinions as well as trends and styles of various periods in American art.

The video section consists of nineteen assorted choices (alphabetized by the artists' names), which accent a variety of media, materials, and artistic traditions ranging from folk art to graphic arts, from video art to the more traditional fine arts of painting and sculpture. Some videos are shot in conventional talking head–style documentary intercut with still photographs and curatorial voice-over, while other videos more directly and subjectively reveal the artist, using the artist's voice and showing his/her working process and work environment. The personal and untraditional content of many of these artist videos counters the otherwise impersonal status of the institution that holds their work. The artist Jon Serl, speaking about other people's artwork, goes so far as to directly address the interviewer/spectator, claiming, "you can do the same damn thing . . . I'm much older . . . you're younger . . . you should be fresher." Serl's provocative and candid statement inserts a playful and personal tone into the institutional sphere, and further offers the user/visitor access to insights otherwise hidden behind the exhibition process. Through their openness, these videos potentially provide the user/visitor with a more comprehensive perspective on art and artists, bringing down the museum walls far beyond a mere reproduction of artwork and re-creation of space.

More than any other title, the Smithsonian's *National Museum of American Art* responds to contemporary museology by allowing the user/visitor a modicum of "behind-the-scenes" access. Seeing the artists on video, hearing their voices and opinions, and watching them work give the viewer/visitor a sense of the production process generally effaced by the traditional museum institution and museum visit. The museum, like most institutions, is structured according to a division between the producers and consumers of knowledge. Tony Bennett has pointed out that this division "assumed an architectural form in the relations between the hidden spaces of the museum, where knowledge was produced and organized in camera, and its public spaces, where knowledge was produced and offered for passive consumption" (1995, 89). In his sociological study of tourism, Dean MacCannell (1976) invokes Erving Goffman (1959) to analyze the touristic experience as, not unlike the museum itself, a highly structured encounter, a distinct and delineated trajectory from front to back regions.[27] Unlike its architecturally inhibited counterpart, the CD-ROM museum, by virtue of its metaphorically constructed spatial structure, can perhaps more easily and provocatively evoke and engage a so-called back region.

The Smithsonian title, aside from access to the artistic production process through the "Media Center," further brings the user/visitor closer to the backstage arena through the "Director's Choice." In this section Elizabeth Broun, the director of the National Museum of American Art, addresses the user/visitor in a short talking head–style video. Despite the conventional documentary format, Broun openly and self-reflexively comments on her selection process, thereby implicitly commenting additionally on the behind-the-scenes selection process of the institution and its curators. She touches on the idiosyncratic nature of selection by admitting that the "Director's Choice" section incorporates famous and historically significant works and lesser-known artists as well as some of her personal favorites from the over 750 works in the CD collection. Broun's comments not only signal her approval of (the user/visitor's) subjective opinion and personal taste; they further encourage the user/visitor to choose and collect his/her own favorites in a "portfolio" (a sort of personal museum).[28] In unveiling a portion of the museum's production and selection process, the Smithsonian title begins to disclose the complicated power dynamics inherent (and otherwise elided) in the visitor-institution relationship.

The only drawback to the Smithsonian's (or Broun's) approach, however, lies in its potential overdetermination of the user/visitor's experience. A less didactic strategy would allow the user/visitor to explore the museum's production, function, and institutional status without such explicit, academic commentary and criticism by a museum guide (curator or director). Such a strategy would allow the user/visitor to uncover the museum through his/her *own* process of discovery. Aside from the conventional museum CD-ROM titles examined above, a few other retail CDs have creatively used the museum environment as a narrative trope or interface. The concept and construction of the museum as a series of rooms enabling a process of discovery, in fact, structure many commercial adventure games that have been produced on CD.

Screen-Play: Demystifying the Museum's Aura

Those alternative museum titles, which more directly appropriate the museum, employ its institutional and spatial constructs to various ends and effects. Through their play, these CDs can suggest ways to more fully meet the new museology's challenges than those that attempt to represent real museums and their collections. While not necessarily addressing the museum directly as a cultural and historical public sphere, these titles playfully expose the museum's constructedness—something that can be appropriated and manipulated. They further explore and exploit the museum's traditional high art status by focusing, and in some cases parodying, the CD-ROM museum's low-culture contents and ephemeral display. Such a playful and parodic approach challenges the elitism associated with the traditional museum as well as the traditional museum CD-ROM collection. These titles therefore offer a newly devised access to and conception of the museum dissociated from the aura of the artwork as well as an aura surrounding the museum institution.

These alternative museum titles include *Eyewitness Virtual Reality Cat* (1996), which constructs a faux natural history museum devoted to the world of felines; *Absolut Museum* (1994), which constructs a series of fanciful gallery spaces featuring 215 of Absolut Vodka's "award-winning" artist-produced advertisements;[29] *Versailles, 1685: Intrigue at the Court of Louis XIV* (1996), which reconstructs a palace as it once existed, filled with mysteries, foibles, and plots to undermine the king; and *Alice* (1994), which constructs a puzzle based on Lewis Carroll's *Alice's Adventures in Wonderland* and *Through the Looking-Glass*, a Japanese artist's (Kuniyoshi Kaneko) two hundred illustrations of the tale, and a rabbit that leads the user on a bizarre journey into a surreal museum of Kaneko's artwork.

The world—or self-titled "interactive museum"—presented on *Alice* is not concerned with re-creating a museum in any conventional sense. Rather, the CD appropriates the spatial dimension of the museum (as well as the spatial and tour-like quality of Carroll's story) as a series of rooms that exhibit arts and artifacts. The user/visitor enters the first room, modeled on the living room in Kaneko's house. Here the user/visitor can explore the space, navigate through the room, and examine its objects (paintings, drawings, books, furniture, knickknacks, etc.). Rules of logic and rules of game play do not govern

Fig. 7.3. The faux museum: *Eyewitness Virtual Reality Cat* uses the museum as a visual frame to uncover a three-dimensional world of felines.

which objects are privileged (those objects the user can click on and thereby activate). The rabbit, thematically appropriated from Carroll's story, subsequently brings the user/visitor into another spatial dimension—a labyrinth of twelve interconnected rooms and hallways. Each room offers a new set of paintings, objects, and keys (or playing cards) to unravel the puzzle. In essence, the user/visitor becomes involved in an elaborate card game to uncover Alice's secrets and find the final door that leads beyond the museum's walls. The objects and paintings play off Carroll's tale, often changing the user/visitor's perspective and size, and further bringing the user/visitor into unknown, parallel dimensions. While the title clearly applies Carroll's narrative, metaphors, and props, it goes beyond the walls of linear storytelling, museum touring, and CD-ROM interfaces to play with narratives, spaces, times, perspectives, and dimensions.

This CD, then, unlike its traditional CD-ROM museum counterparts, offers a platform to explore the concept and construct of the museum. In presenting the museum as an idea, a space, and a part of

Fig. 7.4. The museum playground: the Rabbit leads Alice (and the user) into another dimension.

the game play, *Alice* implicitly challenges the museum's institutional status as well as its site-specific context. The game-like navigation and search process encourage behind-the-scenes exploration and an often unexpected interaction with paintings and objects (a fish jumping out of a dresser drawer, for example). Designed in part by the artist Kaneko and the media performer/director Haruhiko Shono, *Alice* engages the user/visitor as part of Alice's world, and aims to challenge his/her expectations on several levels. Arguably, the alternative CD-ROM museums such as *Alice* parallel the contemporary, worldwide proliferation of alternative and idiosyncratic museum sites.[30] By further undermining the status of the traditional museum institution and celebrating its fetishistic tendencies, these CDs also expand and encourage an exploration of the fundamental relationship and interplay between content and space.

The production of such titles as *Alice* supports Andreas Huyssen's claim that "a museal sensibility seems to be occupying ever larger chunks of everyday culture and experience," and, in turn, "has become a key paradigm of contemporary cultural activities" (1995, 14).[31] The CD-ROM museums not only reflect such a museal sensibility, they likely will engage with and determine the constitution of our ever changing cultural sensibility. It is important, therefore, to recognize and engage the museum and the CD-ROM medium's potential as well as its power. Huyssen's claim may further help to explain and unravel the impact of CD-ROM's existence, use, and conceptual design on future relations between multimedia technologies and the museum as cultural institution. If the museal is in fact occupying more of our everyday life and culture, then its appropriation in various formats and media must be continually and critically acknowledged and examined. The museal sensibility has become part of retail display, theme restaurant chains, advertising, urban gentrification, flea markets, and (as suggested above) CD-ROM games. Its underlying power structures, however, often remain concealed or obscured by their familiarity and everyday accessibility (not to mention their similarity to the culture and display of spectacle that the museum *itself* has adopted). In order to uncover the impact of CD-ROMs and the CD-ROM museum, I have suggested the need to temper the idealistic discourses surrounding the new media. It is equally important, I would argue, to be cognizant of an encroaching museal sensibility and the power structures that accompany it.

As a paradigm of everyday culture, the museum no longer represents a distanced sphere of contemplation or a mausoleum of precious artifacts. The contemporary museum faces a fragmented epicenter and the need to incorporate a wider, more inclusive range of arts and artifacts from high to low, east to west, craft to folk, and so on. While the mingling of high and low cultures is not new to the museum's institutional concerns (and fears), the commercial sphere is increasingly widening its scope, influence, and effect on the nonprofit sector. With these changes, the museum as a cultural and public institution must cater to an ever widening public as both an audience and a financial resource. By virtue of its accessibility, multimedia's admission into the museum can potentially reframe a high-low divide that fuels institutional power dynamics, thereby redressing issues of access. At the same time, however, the presence of multimedia—whether a retail CD-ROM museum reproduction or a resource database used within the exhibition environment—is nonetheless implicated in the museum's constant renegotiation and balance of the high-low divide. As the latest in a long line of mechanical reproduction formats, multimedia has newly altered and displaced the value and concept of exhibition. The historical displacement of art's cult value by its exhibition value, which Benjamin foregrounded in his 1935 essay, must therefore be redefined according to today's standards, exhibition values, and power dynamics.

By stripping the museum of its site specificity and thereby foregrounding it as a social and cultural construct, the CD-ROM museum may illustrate the poststructural critique of the museum more radically than the museum can on its own terms. Clearly, as I have indicated, however, the CD-ROMs rarely meet these challenges. The CD-ROM format, in fact, represents only one of many possible platforms to creatively play with the idea of the museum. In 1994 the J. Paul Getty Museum in Los Angeles designed and produced a retail kit called "Make Your Own Museum." While this kit engages the mechanical over the technological, it offers many possibilities for future ventures between the museum and multimedia technology. The kit consists of six possible cardboard wall designs, six cardboard floor coverings, over seventy works of art from the J. Paul Getty Museum, punch-out figures of a security guard, tour guide, conservator, preparator, and several visitors, and a guidebook, in addition to an array of colorful geometric abstract shapes for creating original artwork. The player is directly invited to participate in the backstage production process by either re-creating the actual galleries of the J. Paul Getty Museum or creating an imaginary museum with any number and variety of objects. The guidebook emphasizes that museum making has no limits, and attempts to encourage originality and personality by suggesting that "a collection can be made up of anything a person finds appealing or interesting, beautiful or mysterious, valuable or fun," and further, that "no matter what your museum ends up looking like, it will tell people a little about who you are and what you are interested in" (Belloli et al., 1994, 8). This overture explicitly challenges museum conventions and values as well as notions of curatorial and institutional authority.

The Getty kit therefore provides yet another way of imagining a museum without walls and limits, whether they be architectural, cultural, or political. This product, like the alternative CD-ROM museums (*Alice, Eyewitness Virtual Reality Cat, Versailles,* and *Absolut Museum*), significantly and provocatively foregrounds the museum as game, learning tool, commodity, artwork, and above all, construct. It demonstrates that the museum can be mastered by the visitor, that the visitor can actively take part in the process of construction, deconstruction, and reconstruction, thereby revising expectations and assumptions tied to a familiar cultural institution. In part, the Getty kit serves as a reminder of the pitfalls of privileging a single technological format, and in demonstrating that new technologies may not offer the only means to challenge old stories and familiar institutions. In addition, like many of the alternative CD-ROMs, the Getty kit confirms the value of looking beyond the walls—the walls of the museum *and* the walls of technology. Although the innovations and uses of multimedia by and within the museum context seemingly open the possibilities of constructing a museum without walls, the power structures behind those walls invariably inform all cultural products and contexts. In order to uncover, or at least be aware of, the power behind these walls, we need to critically examine how new developments in technology will continue to influence cultural spheres, construct new institutional walls, and, in turn, (re)imagine a *musée imaginaire.*

NOTES

1. In her 1986 response to Malraux's essay, Krauss reframes the "museum without walls" in a contemporary postmodern context.

2. Such acceptance of the museum institution's display is comparable to what Stuart Hall (1980) refers to as the dominant-hegemonic decoding or reading experienced by the conventional television viewer.

3. The Corbis Corporation was founded by Bill Gates in 1989 as a privately held company headquartered in Bellevue, Washington. According to Denise Caruso, "although not a Microsoft investment per se, Gates owns (and Microsoft is a customer of) this rapidly growing archive" (1996, 128). The company's mission (according to the company's Web site: www.corbis.com/) is to build a unique and comprehensive archive of high-quality digital and analog images for both consumers and professional image buyers. Professionals can license images from Corbis for use in print and/or electronic media and communications publications.

4. See also Huhtamo (1997).

5. Nash specifically alludes to the backstage spaces (the trailers where the freaks live behind the tent) that contain the "real" stories and "real" identities of the freaks. This exploration of backstage and behind-the-scenes spaces can

serve as an important model for future experiments with the museum and multimedia.

6. For a wide range of critical perspectives on the positive and problematic uses of digital culture in the museum and art historical circles, see Cohen et al. (1997).

7. As many have noted, the access issues pertaining to CD-ROMs necessarily beg questions of class. Therefore, while the museum's institutional elitism may be somewhat diminished by its reproduction and ostensible accessibility through CD-ROM, the elitism of the CD-ROM technology itself (along with personal computer ownership/access) still remains a significant factor in any discussion of access.

8. This was a common sentiment among CD-ROM producers, museum educators, and electronic publishing executives interviewed for this essay.

9. For a general discussion on the influence of commercialism, entertainment, and spectacle on the museum, see Lumley (1988), Sherman (1994), Macdonald and Silverstone (1990), and Harris (1990).

10. Most of these simulations/recordings employ QuickTime®.

11. Most museum curators and administrators firmly discount the notion that a virtual visit through CD-ROM or the Web could ever supplant the real visit, the physical entry of a space. It remains to be seen whether new generations raised on computers and digital imaging will agree.

12. Over the next few years, the level of museum activity and involvement in multimedia will undoubtedly begin to change. The use of the Internet will be a significant factor. Rather than focusing on the retail market, the average museum will likely concentrate its outreach into more specific arenas such as schools, universities, and libraries. Additionally, many museums have been increasingly employing the CD storage platform in the development of touch-based kiosks located in the museum. With its extensive financial resources and commitment to education, community development, and historic preservation, the new Getty Center will undoubtedly serve as a model for the broad potential (as well as the limitations) of new technologies in the museum environment.

13. For a list of museum Web sites, see Noack (1995) or visit the Guide to Museums and Cultural Resources at http://www.usc.edu/lacnmh/webmuseums/.

14. Besides book stores, which are increasingly pulling out of the CD-ROM marketplace, specialty CD-ROM stores are more likely to carry fine art and more experimental titles (by companies like Voyager, for example). Other sales venues for such titles include warehouse sales, direct sales, and museum stores. National chains like Software Etc. and Egghead Software are less likely to carry such titles.

15. Representatives from Digital Arts and Sciences Corporation and Mac-

millan Digital have supplied this opinion, suggesting the overall difficulties of sales in the fine art CD marketplace. The realities of production expenses, labor costs, and market return have made many companies rethink their investment in CD-ROM as a viable source of revenue. Unlike other media industries, such as film and television, the CD-ROM industry lacks a stable market base on which analysts can predict and evaluate its sales and future success rate. Corbis's claims to success, therefore, may be problematic in terms of predicting future success in sales and distribution. The company's success may be attributed in large part to the quality of products it develops. These products tend to be more entertainment than reference-oriented, often featuring attractive components such as navigable spaces and filmic introductions and interludes. In addition, the company tends to foreground the biographical narrative over the institutional in titles ranging from *Passion for Art* to *Cézanne* and *Leonardo da Vinci*. The company's success, however, may also be attributed to its production and distribution approach. Gates's company tends to be more solidified and streamlined in terms of its production process, whereas most other companies tend to operate in a more fragmented field of development, production, publishing, distribution, and marketing where work is contracted out to smaller, independent companies.

16. Digital Arts and Sciences Corporation is at the forefront of developing and distributing innovative imaging and image-based asset management software designed specifically for museums. In 1993 the company took on a three-year legacy of product development as a technology spin-off of AXS (software products combining digital images and text for study, research, and entertainment). The company subsequently produced its own software product, EmbARK, to provide a new standard for museum collections and management systems. This product "pioneered the transition to image-based databases" (DAS Web site). In 1996 the company acquired Gallery Systems of New York, whose collection management system, the Museum System, is installed at the Metropolitan Museum of Art and other institutions worldwide. Behind-the-scenes software systems such as these may also shape what is used in museum resource systems and on museum Web sites.

17. According to information on the Corbis Corporation Web site, nearly one million of these seventeen million images are in digital format, making it the largest collection of digital images in the world. Thousands of new images are added each week. The archive currently works with the following museums: the Barnes Foundation, the National Gallery in London, the Philadelphia Museum of Art, the State Hermitage Museum, the Kimbell Art Museum, and the Seattle Art Museum. The archive also holds several premier photography collections, including the Bettmann Collection, the LGI Collection, and the Turnley Collection, as well as the extensive work of Ansel Adams and Roger Ressmeyer.

18. For a more complete discussion of legal access and fair use with regard to art and digital culture, see Cohen et al. (1997).

19. For example, see Dean (1994). This "textbook" for curators and museum designers methodically details the requisite approaches to planning, developing concepts and story lines, targeting audiences, designing spaces, constructing and controlling the exhibition environment, and so on. In one section, Dean draws on behavioral science to make claims about visitor tendencies and movements through space (e.g., the tendency for most people to turn to the right if all other factors are equal). In another chapter, "Storyline and Text Development," Dean outlines the key elements of the story line "process," including narrative document, outline of exhibition, list of titles, subtitles, and text, and list of collection of objects. This comprehensive, though elementary, "how-to" text manifests the institution's powerful narrative traditions and formulas.

20. Costs for constructing a navigable space for the museum CD titles are considered by many to be a waste of time and money. One publisher cited *A Passion for Art* as a particular example of an overextended budget. This suggestion was denied by the Corbis Corporation, however, which considers *A Passion for Art* to be its best-selling title; it has sold well for a number of years in the retail market.

21. While the use of black and white in these cases is most likely a decision based on program memory, it also significantly reflects on authenticity issues pertaining to reproduction.

22. The level of interaction between user and material may be somewhat inhibited, however, by the impersonal status of the user's identity as a museum visitor. The user's identification is delimited by the cursor. In addition, the user's experience of the museum environment, unlike a real visitor's, is completely solitary. There are no other visitors, staff, or museum guards. The user therefore lacks a degree of interpersonal contact and interaction that is very much a part of the real museum environment.

23. While Gunning's discussion of the "cinema of attractions" in many ways parallels the early developments and attractions of CD-ROM technology on the level of spectacle and spectatorship, there may also be significant differences worth exploring. Gunning clearly privileges early cinema's active spectator who participates in the cinematic experience like a heckler at a vaudeville show. The CD-ROM similarly privileges an (inter)active spectator/user. However, the solitary nature of CD-ROM use significantly alters the nature of this (inter)activity at the level of participatory (or audience) experience.

24. Gunning (1990) has argued that we need to examine the early period of cinema as a field in its own right rather than dismissing or discounting it as primitive. Similarly, the CD-ROM should be examined in its own right

through a system or model that can adequately manifest the degree of difference and range of play in its textual production and consumption. Gunning's suggestion of a generic system of analysis for early film may serve as a useful and analogous model for the CD-ROM medium and the CD-ROM museum.

25. Many have criticized the crudeness of CD-ROMs that depend on spatial navigation. The speed and accessibility of movement through space have improved significantly; however, a more "virtual" navigation platform is still not readily available in the consumer marketplace. The game industry tends to motivate and drive such technology. Nintendo 64's use of processing power may lead the way in new approaches to real time–3-D. However, many in the industry are also waiting to see what develops with the coming convergence of television and the Internet.

26. This is especially the case for the archive section of the Barnes Collection. The CD-ROM provides digitized copies of documents lost until 1989. According to the CD's introductory guide, these documents and audio sound bytes had never before been displayed to the public.

27. See also Meyrowitz (1985).

28. The user-defined "portfolios" contain up to twenty-five works of art or clipped text that can be saved, printed, personally annotated, or displayed. The Smithsonian title also offers a "Datadisc" section where users can locate and display additional artwork in the collection that is similar in style or medium to personally selected works of interest. It may be problematic, however, to celebrate the self-reflexive tone and backstage access offered by these sections; a certain degree of cultural capital may be required for the user/visitor to directly and productively engage them, and therefore a utopian notion of equalized access is further complicated and premature.

29. *Absolut Museum* is currently produced on three floppy discs rather than CD-ROM. My inclusion of this product/title in an essay on CD-ROMs indicates the dilemma behind elevating one storage format (floppy disc, CD, DVD, etc.) over another. Despite its distribution on floppy, *Absolut Museum* permits a similar, if not more enhanced, degree of fluid spatial navigation comparable to the Corbis titles.

30. See Rugoff (1995) for a discussion of alternative museums in America's changing visual landscape. Among other sites, Rugoff discusses the Museum of Jurassic Technology, the Los Angeles County Sheriff's Museum, and the Liberace Museum. Other "alternative" museums collect items such as shoes, lunch boxes, frozen food, barbed wire, neon lights, and toilet tissue.

31. While not referring directly to the CD-ROM, Huyssen offers the "electronic totalization of the world on data banks" as exemplary of a museal sensibility. Add to this the plethora of theme restaurant chains such as Planet Hollywood, the Hard Rock Café, and the All-Star Café as well as home shopping catalogs that feature museum-like display cases for home, rewriting

the collection mentality associated with the cabinet of curiosities for today's more mundane, everyday collectibles.

WORKS CITED

Belloli, Andrea P. A., and Keith Godard. 1994. *Make Your Own Museum*. New York: Ticknor and Fields Books for Young Readers; Malibu: J. Paul Getty Museum.

Bennett, Tony. 1995. *The Birth of the Museum: History, Theory, Politics*. New York: Routledge.

Bourdieu, Pierre, and Alain Darbel. 1990. *The Love of Art: European Museums and Their Public*. Trans. Caroline Beattie and Nick Merriman. Stanford: Stanford University Press.

Caruso, Denise. 1996. Microsoft Morphs into a Media Company. *Wired*, June, 128.

Cohen, Kathleen, James Elkins, Marilyn Aronberg Lavin, Nancy Macko, Gary Schwartz, Susan L. Siegfried, and Barbara Maria Stafford. 1997. Digital Culture and Practices of Art and Art History. *Art Bulletin* 79, no. 2 (June): 187–216.

Dean, David. 1994. *Museum Exhibition: Theory and Practice*. New York: Routledge.

Donato, Eugenio. 1979. The Museum's Furnace: Notes toward a Contextual Reading of *Bouvard and Pécuchet*. In *Textual Strategies: Perspectives in Post-Structuralist Criticism*, ed. Josué V. Harari, 213–38. Ithaca: Cornell University Press.

Edson, Gary. 1995. *International Directory of Museum Training*. New York: Routledge.

Goffman, Erving. 1959. *The Presentation of Self in Everyday Life*. Garden City, NY: Doubleday.

Gunning, Tom. 1986. The Cinema of Attraction(s): Early Film, Its Spectator and the Avant Garde. *Wide Angle* 8, nos. 3–4: 63–70.

———. 1990. Non-Continuity, Continuity, Discontinuity: A Theory of Genres in Early Films. In *Early Cinema: Space, Frame, Narrative*, ed. Thomas Elsaesser with Adam Barker, 86–94. London: BFI.

Hall, Stuart. 1980. Encoding/Decoding. In *Culture, Media, Language*, ed. Stuart Hall et al., 128–37. London: Hutchinson.

Harris, Neil. 1990. *Cultural Excursions: Marketing Appetites and Cultural Tastes in Modern America*. Chicago: University of Chicago Press.

Huhtamo, Erkki. 1996. Digitalian Treasures, or Glimpses of Art on the CD-ROM Frontier. In *Clicking In: Hot Links to a Digital Culture*, ed. Lynn Hershman Leeson. Seattle: Bay Press.

———. 1997. Most Multimedia Sucks: An Email Exchange with Michael Nash. *Atlantic Unbound*, 5 June (http://www.TheAtlantic.com/).

Huyssen, Andreas. 1995. *Twilight Memories: Marking Time in a Culture of Amnesia*. New York: Routledge.

Krauss, Rosalind. 1996. Postmodernism's Museum without Walls. In *Thinking about Exhibitions*, ed. Ressa Greenberg, Bruce W. Ferguson, and Sandy Narine. New York: Routledge.

Le Corbusier. 1987. *The Decorative Art of Today*. Trans. James Dunnett. Cambridge: MIT Press.

Lumley, Robert, ed. 1988. *The Museum Time Machine: Putting Cultures on Display*. London: Routledge.

MacCannell, Dean. 1976. *The Tourist: A New Theory of the Leisure Class*. New York: Schocken Books.

Macdonald, George F. 1992. Change and Challenge: Museums in the Information Society. In *Museums and Communities: The Politics of Public Culture*, ed. Ivan Karp, Christine Mullen Kreamer, and Steven Levine. Washington, DC: Smithsonian.

Macdonald, Sharon, and Roger Silverstone. 1990. Rewriting the Museums' Fictions: Taxonomies, Stories, Readers. *Cultural Studies* 4, no. 2 (May): 176–91.

Malraux, André. 1990. *The Voices of Silence*. Trans. Stuart Gilbert. Princeton: Princeton University Press.

Meyrowitz, Joshua. 1985. *No Sense of Place: The Impact of Electronic Media on Social Behavior*. New York: Oxford University Press.

Nash, Michael. 1996. Vision after Television: Technocultural Convergence, Hypermedia, and the New Media Arts Field. In *Resolutions: Contemporary Video Practices*, ed. Michael Renov and Erika Suderburg. Minneapolis: University of Minnesota Press.

Noack, David R. 1995. Visiting Museums Virtually. *Internet World*, October, 86–91.

Rugoff, Ralph. 1995. *Circus Americanus*. London: Verso.

Serl, Jon. 1996. "On Other People's Art," video interview on *National Museum of American Art, Smithsonian Institution*. New York: Macmillan Digital USA.

Sherman, Daniel. 1994. Quatremère/Benjamin/Marx: Art Museums, Aura and Commodity Fetishism. In *Museum Culture: Histories, Discourses, Spectacles*, ed. Daniel Sherman and Irit Rogoff. Minneapolis: University of Minnesota Press.

Eight

Virtual Kinship in a Postmodern World
Computer-Mediated Genealogy Communities

Pamela Wilson

Since the mid-1990s a convergence of technological innovations and trends—the increased accessibility of personal computers, the explosion of the genealogy software market (now primarily in CD-ROM form), and increased access of the general public to the Internet—has contributed to the growth, especially in America, of an extensive and interconnected community of amateur historians who rely on computer technologies to uncover the secrets of their personal family histories. They constitute the little-acknowledged but increasingly popular practice of historical research being done outside the realm of professional or academic historiography. This historical practice, deeply inscribed in new computer-mediated technologies and industries, has effected new types of discourses and communities based around ancient tropes of family, kinship, and mutual histories, yet deeply rooted, both economically and culturally, in what might be termed the "postmodern condition."

Background: Genealogy, Privilege, and Whiteness

In my introductory social anthropology class at an American liberal arts college in the mid-1970s, I had learned about the importance of kinship structures as the basic system of social organization in tribal societies, and was taught the various systems used by Western anthropologists for coding and charting descent and lineage in these exotic societies, so different from our own "Western" culture. Sociologists of the time spoke (regretfully) of the social and familial alienation inherent in contemporary modern civilization, and sociocultural anthropologists were only then beginning to conceptualize the implications of kinship and friendship structures in Euro-American societies.[1]

However, what my professors didn't teach me—perhaps because it fell through the disciplinary cracks between anthropology, sociology, and history, among others—was that genealogy (the study of one's ancestry or family history) had been around a long time in Western European society, and had acquired a traditional connotation as a class-based desire to prove a genetic pedigree in order to acquire social privileges. Problems of classism still plague genealogy today and relegate it to an amateur and nonacademic field of historical and social research. Although a particularly American approach to genealogy has acquired its own flavor, especially in the last quarter century since the 1960s, it stemmed originally from an Anglo-European endeavor to define and delineate those with social privileges as distinct from those without. The two basic defining possessions in Europe were "blood" (genetic claims) and landed title. As Wagner explains,

> The ruling classes in the sixteenth and seventeenth centuries took a keen interest in their pedigrees. Modern scientific genealogy, indeed, originated in this period. It was a sophisticated interest, linked with family pride and a love of history, and catered for by professional genealogists. . . . During this period it came to be widely thought that pedigrees were proper only to distinguished families, to the owners of lands and titles and their progeny. . . . The heralds, antiquaries and county historians were happy to include the pedigrees of minor landowners, wealthy merchants and professional men in their collections. But the assumption [held] that a pedigree should be linked with an actual or potential stake in the country. (Wagner 1960:2)

In British genealogy even today, there are tomes providing the peerages, baronages, and other family histories of the landed gentry, important sources of documentation in a society in which one branch of the ruling parliament is a hereditary House of Lords. There is also a huge amount of European genealogical literature on the ancestry of prominent and royal figures, such as kings, princes, counts, and other political leaders, dating back to the early Christian era and even before. American genealogical researchers connect into this vein of research when and if they happen to trace their immigrant ancestry back into the intertwined networks of British aristocracy and royalty in the late Middle Ages—a network from which many of the so-called gateway ancestors of colonial America were descended.

In the United States, then, institutionalized genealogical research from the colonial period until the mid-twentieth century centered primarily around this "founding fathers" mentality, usually in attempts to prove the connections of socially and politically prominent American families to the established pedigrees of European nobility.[2] A second type of American genealogical impulse, no less aristocratic (but less European-oriented), reflected the pride and prestige associated with being descended from select American pioneers and patriots—thus leading to the establishment of associations of Mayflower descendants, the Social Registers of various northeastern cities, the Daughters of the American Revolution, descendants of Colonial Virginia's Pioneers and Cavaliers, and so on.

It should be evident to even the casual observer that, owing to the Anglo-European origins of all of these groups, the demographics of traditional genealogy in Europe and America have been mostly white; in fact, one might argue that, as an institutionalized practice in Anglo-American society, genealogy encodes and celebrates whiteness and white privilege. Although not overtly racist in discourse, the element of genetic distinction and prestige associated with claims to pedigree reinforces a social structure that has traditionally and institutionally effaced ethnic and racial otherness and reinforced hereditary white privilege.

However, since the changes in American society in the 1960s, and especially since the 1976 publication of Alex Haley's immensely popular *Roots* celebrating African American heritage, there has been a palpable change in the feeling surrounding American genealogy as a practice. This gradual shift in the 1970s and 1980s has become a fast-moving transformation in the 1990s, with the addition of computer-mediated and online resources to foster the explosion of popular interest in amateur genealogical research. Another major force, however, that has helped to popularize genealogical research as a middle-class hobby but also reinforced the racially and politically conservative reputation of the practice has been the Church of Jesus Christ of Latter-Day Saints (Mormon). The Mormon Church is a born-in-America religion that has taken a leading role in institutionalizing, and ultimately industrializing, the process of genealogical research because of its religious rituals, which center around the posthumous baptism of believers' ancestors. Its Family History Division is a major provider of archival resources, research library facilities, and the archetypal CD-ROM research database. Yet the Mormon influence on genealogical research has reinforced its perception as primarily a white, middle-class, and politically conservative avocation.

According to a 1995 article in *American Demographics* based on marketing surveys, genealogy in the 1990s is one of the most popular hobbies in America; 40 percent of American adults are "somewhat interested" in pursuing their family history, and 7 percent claim to be "deeply involved" in the practice. Interest in genealogy seems to be popular across all age ranges and socioeconomic levels, though there is a concentration of most intense involvement in the hobby among those between thirty-five and fifty-four, and with annual incomes above $55,000 (Fulkerson 1995).

In spite of the demographic skewing toward aging white baby boomers with disposable income, there is also a growing interest in genealogical research among America's ethnic minorities. The practice of tracing their family history in the Americas has been made difficult for African Americans, especially those whose ancestors' identities were effaced and discounted through the practice of slavery. Yet the challenge of piecing together those lost generations, incited by Haley's book, has been met by an increasing number of African Americans, whose presence is increasing in historical archives and on the Internet. Although a part of the general online genealogical community, each ethnic or racial group has distinct networks of sources and fellow researchers to aid them in their searches, and organizations like the African American Historical and Genealogical Society have made specialized publications and resources available.[3] In light of the presence of other specialized resources, genealogical research is apparently also thriving as a hobby among Americans of Hispanic/Latino, Jewish, Italian, Slavic, West Indian, and Native American descent.[4] Thus one might characterize those who are interested in, and actively pursuing, their personal and family histories in the late 1990s as a cross section of the American public, residing in nearly every habitus of American society. However, their motivations for engaging in the searching behaviors that lead them to do research in libraries, buy the CD-ROMs, and participate in the Internet-facilitated communities of researchers vary tremendously.

Although genealogical research itself is certainly not new, these innovative modes of research via online and CD-ROM sources have given birth to new modes of participating in family history research. In particular, the 1990s have ushered in new models and systems for recording and managing research data, and for publishing, distributing, and sharing the results of one's research with the wider public. The computer-mediated amateur genealogy "industry" has quickly transformed "genealogy" from a localized, pen-and-paper endeavor—of amateur, personally published, limited-edition books—to participation in the market economy of major software companies like Brøderbund and Parsons Technologies. The means of production of "doing genealogy" has been radically transformed—including the very quality and method of research (data gathering, record keeping, and organization of specialized knowledges) as well as the style, mode, and content of publication and the mode and scope of distribution (dissemination of specialized knowledges). Even the audience for such knowledges has radically changed and widened. What was once an extremely localized network for distribution of very specialized materials, seemingly of very narrow interest (for example, a researcher might have printed and distributed only a few hundred copies of a specific family's history), has now been broadened through the global reach of the Internet and the marketing of the software companies to locate and create an audience community whose individual members are dispersed around the world.

Computer-related technologies have, therefore, radically reshaped the localized cottage industry of genealogical research/publication into a truly hybrid commercial/noncommercial postmodern industry. Inscribed within this industry are a multitude of commercial and noncommercial "players" who represent a number of diverse (and often conflicting) interests. Though interdependent, these interests are often in ideological and discursive conflict, as we shall see, about whether historical knowledge is public or private, able to be packaged, bought, and sold. On the commercial end we find the major software publishing companies, led by Brøderbund, which have provided invaluable tools to enable amateur genealogists to do their research. On the other end, and along a broad spectrum, we find a vast and loose network of noncommercial organizations and individual initiatives that use the Internet to create a public sphere of common discourse. Their research is aided by civic and academic institutions housing historical and archival materials, whose collections are increasingly being digitized (such as university and state libraries), and the Family History Division of the Mormon Church, an enigmatic but central institutional figure in the transformation and industrialization of genealogy.

Genealogy as a Discursive Practice

For many people the thought of genealogy dredges up images of pages upon pages of yellowed charts of names, written in the shaky but exacting penmanship of an elderly great-aunt nearing her deathbed—or arouses a sense of boredom with the minutiae of history-as-divorced-from-personal-relevance, as Geoffrey Ward recently explained:

> Anyone who has endured an evening with an overly enthusiastic amateur genealogist knows how truly tedious family history can be: the sheer volume of faceless names and dates, the tenuous links to the celebrated or notorious, the infinite permutations of cousinhood. To such people all history seems to exist only to fill out branches of the family tree. (Ward 1995: 16–17)

Why *do* people find genealogy interesting to study? Why does genealogy seem so fascinating to some and so loathefully boring to others?

One of my correspondents proposed, only somewhat facetiously, a "genealogy gene" that might predispose some of us to historical inquiry:

> Long ago, when my first child was born, there was a place in her baby book for ancestors, and I didn't know some of the names needed; it came as a shock to me. I hustled about and found the necessary names and tried to forget the subject, but there must be a genealogy gene. I always listened and remembered the family stories my parents told. (My sisters did not and they never were interested in genealogy.) I am a story teller and these true stories of ancestors were repeated to my children and grandchildren.
>
> I don't think we can say exactly why we are driven, but I feel a real kinship with the ancestors I find once I know a bit about their lives. When my grandson was 8, he said to me, "I wish Stephen Hopkins (Mayflower) hadn't died. I have some questions I would like to ask him." Does my grandchild feel a connection to Hopkins that many

> descendants do not feel? Is this the genealogy gene at work? I have never been able to generate real interest from some cousins and siblings, but others just fall into step. I think we may be born with something just a bit different in our makeup if we are truly dedicated researchers. I have no interest in having the longest list of names with connections back to the dark ages, but I want to make each person in my ancestry become *real* and not just a name on paper.[5]

Other enthusiasts are drawn to the hobby of genealogical research as compulsively as some are pulled into practices of fandom and related behaviors. It becomes an activity that engages us at many levels, activating our desire to find out more and more about a topic, to obsessively complete an implicitly infinite chart with blanks just waiting to be filled—like a crossword puzzle—or to participate with other like-minded and enthusiastic folks in a large party game that taps a very personal vein as well as demanding considerable intellectual and problem-solving skills.

In order to understand the fascination with genealogical research, as well as the ways new electronic technologies like CD-ROMs and the computer-mediated information networks have affected—and in fact transformed—the practice of doing genealogical research, I became deeply engaged in this amateur genealogical community as a participant and ethnographer during the spring of 1997. I was little prepared for the enthusiasm and overwhelming intensity of both historical research and interpersonal interaction that awaited me as a member of this community, or the sheer amount of historical information that I would find circulating on CD-ROMs, on computer discs, and through electronic mail. I became an active consumer and participant in the CD-ROM marketplace (which entails not only purchasing, but also contributing to, the publishers' CD-ROM databases). I pursued the leads and clues I found in the CD-ROMs with the zeal of a detective, which led me into informational exchanges with people on several continents by way of e-mail, the postal services, and many hours of long-distance phone calls.

In addition to the obvious consumer economy into which users of commercial CD-ROMs are inscribed, I soon learned that participation in the genealogical community creates an elaborate interactive and noncommercial *economy of knowledge exchange*, fascinating in its complexity. In order to participate in the economy, one must first gather as much specialized knowledge as possible, using all available resources. Then, to be accepted as a full-fledged member of the community, one must be *both* a receiver and a generous contributor in the circulation and exchange of genealogical knowledge. The level of generosity in this purely noncommercial level of exchange (which is markedly distinct from the commercial economy of the CD-ROM and software industries) is truly astounding. Fragments of genealogical knowledge—the currency in this economy—came to me via e-mail attachments and specialized computer files, handwritten documents and photocopied archival materials arriving via Priority Mail, and even computer discs with manuscripts of family history books provided by generous authors eager to share their research. I joined and became actively involved in a number of online discussion groups and mailing lists. In the end (though there is really no end, for the searches always continue), I extended my sense of my personal kinship circle—people who share a common ancestor with me—at least a thousandfold, and count a handful as respected colleagues who have collaborated closely with me on numerous detective missions.

During the course of this ethnographic research I discovered a world of passionate, nonacademic historians, many of whom are racing to take advantage of every innovation computer technologies can lend them. In doing so they are being met head-on by the software publishing companies, eager to capitalize on this growing trend. To make matters more complex, many other institutional players have staked interests in amateur genealogy. This union of wary but pragmatic bedfellows has engendered a complex dynamic of tensions about old versus new technologies, about the commercialization of historical knowledge, and especially about whether the Internet (and other related venues for electronically publishing such information) should be a public, noncommercial space or a commercialized market space.

Public Space versus Market Space: The Business of CD-ROM Publishing

It is really not possible to discuss computer-related genealogy research without including all aspects of the amateur "industry," since the CD-ROM/software industry and its users, the Internet organizations, databases, and Usenet groups, the e-mail connections (mailing lists and private correspondence), and other modes of interpersonal correspondence are all so intimately interconnected.

First, almost everyone who "does" computer genealogy uses some sort of software package for personal research (informational storage and data organization), so the basic market for the software companies is closely tied to the increasing numbers of hobbyists who are discovering genealogy on the Internet. These packages are almost all built on a computer file format system called GEDCOM, developed in the 1970s by the Family History Division of the Church of Jesus Christ of Latter-Day Saints, and this GEDCOM format serves as the basic format for translation among all the different commercial software programs, such as Family Tree Maker, Family Origins, Brother's Keeper, The Master Genealogist, or Family Gathering.

Second, the software companies maintain a strong and central presence on the Internet, and perceive the integrated CD-ROM/Internet informational system as vital for their operation and marketing strategies. The major genealogy software company, Brøderbund (which produces Family Tree Maker and which, since August 1997, also owns rival Parsons Technology's Family Origins software), has an extensive Internet presence designed to support its software users, attract new users, and market its constant production of new supplemental software. Brøderbund's Web page boasts what is perhaps the largest and most detailed search mechanism on the Internet for searching by an ancestor's full name (as opposed to the more common surname searches). What a searcher finds with the Family Tree Maker "Family Finder Index," however, is a list of Brøderbund-published CD-ROMs on which that name appears—and all are available for order at the click of the mouse and with a credit card number.

Various publishing companies, ranging from the major software companies to minor publishers specializing in genealogical books, have begun to market CD-ROM databases, usually in the form of searchable reference collections of primary sources. These include important public domain records (public documents such as census, birth, marriage, and death records). Publishing companies are also putting primary research books and guides onto CD-ROMs, as well as back issues of genealogical journals (e.g., the *New England Genealogical and Historical Register*). All of these are basic reference materials that are available in either public records offices or public libraries, but which until recently have been very difficult to access outside local archives or specialized genealogical library collections. Many of the now-digitized collections of public documents had been transcribed from handwritten original sources by local volunteers, as part of the efforts of local and county historical societies over the past thirty to forty years, and were generally published first in limited editions by specialized genealogical publishing companies before being recently republished as parts of CD-ROM collections. This is one example (of which there are many) of the extreme personal labor with little reimbursement that has characterized genealogical historiography, and continues to do so even in the age of the Internet.

A second type of CD-ROM database, and one that is growing in popularity though not without its detractors, consists of secondary sources—family trees (sometimes called pedigrees) and family histories that have been collected from various professional and amateur researchers. These databases range from electronically republished family history books (usually those originally published by small specialized publishers and with little chance of a wide national or international distribution before the advent of CD-ROMs) to family trees submitted by consumers and contributors. The latter represent a booming trend, led by Brøderbund's World Family Tree project and the noncommercial Mormon Ancestral File project, which has aimed to create large databases of already researched family trees (interconnected and consolidated ones, in the case of Ancestral File), so that novice researchers can perhaps find some branches of their family history already researched. An important distinction may also be made between the convenient, home-based commercial CD-ROMs, such as the Brøderbund series, and the institutional Mormon series, which were the original prototypes for genealogical CD-ROMs but which are restricted to use only in Mormon Family History centers.[6]

The detractors of these databases compiled from consumer submissions, of course, argue that one cannot trust the validity of these sources, since they are in many cases poorly researched, poorly documented, and so on. Some argue that the credibility of genealogical researchers is lowered with such "cutting and pasting" strategies of procuring information:

> Right now, many folks are busy swapping records and not doing actual research. . . . I have done a bit of this record swapping myself, but I know the difference between getting records from someone who has already done the research and doing it myself. I fear there are many

> beginning genealogists who haven't the foggiest idea of how to do actual research.[7]

The most strident opposition to the World Family Tree project, however, reflects the inherent tension between the commercial and noncommercial impulses of computer genealogy. Opponents argue that by submitting the results of one's research to Brøderbund, which has amassed enough files to create thirteen World Family Tree CD-ROMs at this writing (each of which sells for $39.95), the researchers are contributing to Brøderbund's profit but getting nothing in return. Many of these people would rather make their family trees available at no cost on their own Web sites than have them sold as part of a corporate package. Others want more careful control over their productions, and prefer to find ways to publish books either traditionally or electronically that will include more, textually, than just the basic family tree information that Brøderbund's product provides (e.g., stories, photos, or documents).

How can computers enhance family history research? What can one do with a computer? And how is it different from "traditional" ways of doing genealogical research? Members of the genealogical community who responded to my online questions shared some of their experiences:

> Record keeping and sharing is simplified. I find the idea of a small handful of disks holding as much information as the many file drawers I once used delightful. Corrections are easily made and additions are a snap. Making prints or GEDCOM files is simple, so record keeping is far superior with a computer than it was with a typewriter. I have seen much progress with source material just this past year.[8]

> I have a genealogy program for my PC which I've entered about 1500 names into so far. I will have to go back eventually and enter all my notes and sources, but in the meanwhile, I now have the capability of printing out charts and family group sheets. This has been a real help with getting myself to correspond with relatives. Before, it was just too time-consuming to fill out a chart and then photocopy it as well as write a letter. Now, I've begun to really compare notes with other interested relatives.
>
> My program came with a couple of CD-ROMS of the Social Security Death Index, and also with two CD-ROMS of Brøderbund's "World

> Family Tree." I have used the SSDI to get death dates for cousins who lived until recently. From there, I have been able to get a few obituaries that gave more family information. In one case, I tracked down a second cousin by using the obituary of her mother, and re-discovered one branch of my grandmother's family tree.[9]

In discourses about genealogical method, a topic that is hotly debated in the genealogical community, there is a clash between the more traditional and time-consuming library-based paper-and-microfilm chases and the high-tech database searches of CD-ROMs and Web sites:

> Which is better—the Old-fashioned Detective Work or Research with a Computer? As a whole, the old fashioned way of just researching from yourself backwards has produced more information for me. . . . For years I just picked the brains of all of my relatives, haunted cemeteries in the towns where I located ancestors and went through reams of microfilm of newspapers. . . . Then in 1990, we moved [away from] my five libraries, my county courthouse, my cemeteries, and most of all, my genealogy societies. So I bought a computer after I heard about modems. A whole new exciting world opened up to me, and after three years of learning how to use them and inputting genealogical data, I'm now on my second computer.[10]

Just as in fields of academic research, there are the concerns about the potential proliferation of inaccurate, biased, and/or poorly researched information, which is often accepted uncritically by the novice researcher. A lot of genealogical material that circulates on the Internet as well as on the mostly unverified consumer-submitted CD-ROMs (such as the Brøderbund World Family Tree series) is, in the words of one researcher, "just plain wrong." For serious amateur and professional genealogists, historical truth—proved, verified, and documented with acceptable evidence—is everything.[11] Since much of family history is folkloric in nature—and many of us place a high value on those family stories and legends—there is a degree of tension and skepticism between those who advocate the more detailed, "just the facts, ma'am," style of doing genealogy and those who paint a colorful canvas of family and social history with broader (and perhaps less accurate) strokes. A tension between "old ways" and "new ways" of doing genealogy is often expressed, and may also reflect some major generational differences in the practice of genealogical research that go beyond, yet often map onto, the other historical changes in attitudes about who "does" genealogy, why they do it, how they go about it, and how they publish it.

Genealogy on the Internet

As mentioned earlier, the intimate interdependency of the Internet resources for genealogy and the CD-ROM/software industry fosters a dialectic of both support and competition between the two modes of computer-related research. There is a strong dependency of the commercial publishers on the online community (and vice versa), but also an ongoing tension about the ultimate commercial (market space) or noncommercial (public space) nature of the Internet, specifically with regard to the packaging and selling of public domain knowledge by software companies.

The genealogical information available on the Internet takes many forms, ranging from the commercial to the noncommercial, from the very personal to the institutional, and diverging in many other ways in terms of style and structure. The most commercial sites are those linked to the CD-ROM publishing companies. For example, Brøderbund's Family Tree Maker Online site[12] has established itself as a no-miss site for any genealogical researcher, and boasts that it is one of *PC Magazine*'s top hundred Web sites. Formatted like a magazine or newsletter, it contains how-to articles on different aspects of genealogy and provides the Family Finder Index (mentioned earlier, which indexes online all of the Broderbund genealogy CD-ROMs by the full names of individuals contained therein), message boards and classifieds for members (FTM software users), ordering information for its various CD-ROM products, and some for-fee services such as record lookups and genealogical research associates for professional research. Brøderbund also offers genealogy home page sites to its members, and has a search function that visitors can use to search the user home pages by names of ancestors. All of Family Tree Maker Online, then, functions to provide support services to its software users and to motivate newcomers to use its software and purchase from its extensive CD-ROM collection. In short, the Web site serves as a marketing strategy for the CD-ROM products.

Alternatives to the commercial institutional sites on the Internet are the noncommercial sites that have put genealogical and historical resources online. These include online genealogical societies as well as online archival and library collections ranging from university libraries and state archive collections to localized family and local history centers. These archival materials are becoming increasingly accessible online, as institutions like state libraries and historical societies acquire funding and personnel to digitize their archival documents.

Of a very different nature are the volunteer-organized cooperative genealogical projects dedicated to providing free and noncommercial historical material to genealogical researchers. One example is the ambitious U.S. GenWeb Project, a massive attempt to impose order on the inherent chaos of genealogical information on the Internet.[13] This was a noncommercial enterprise started by a group of Kentucky genealogists in 1996:

> The idea was to provide a single entry point for all counties in Kentucky, where genealogical data about each county could easily be found. In addition, the data on all county sites would be indexed and cross-linked, so that a single search in the master index could locate all references to a given surname across all pages and databases associated with the project. The discussion quickly turned into a reality, and the project snowballed so quickly that by mid-July (1996) all the Kentucky counties had pages. [At this time] it was decided to extend the concept to all of the United States.[14]

Efforts followed to consolidate the localized historical information, gathered and archived at the county and state level, that might be useful for online genealogical research. The project is run by a network of volunteer administrators and coordinators, organized by state and county throughout the United States, who create and maintain home pages, host mailing lists and discussion groups, and transcribe public domain records (census records, marriage bonds, property records, etc.) from local archives for online access.

Perhaps the most significant noncommercial integrated database and informational enterprise is the RootsWeb Genealogical Data Cooperative, home for a decade to the pioneering genealogical discussion list and the mother of all searchable surname databases, a registry that lists and cross-indexes the surnames being researched by all people who choose to submit their research interests.[15] This is one of the first "places" people go when they begin researching a family name by computer. The RootsWeb project has two missions: to provide access to large volumes of genealogical data, and to provide support services to online genealogical activities by serving as home to hundreds of genealogy-related news groups and mailing lists. As their literature explains, "We created the RootsWeb site to continue to support the genealogical community as the needs of the community outstripped the ability of . . . other organizations to host the genealogical community on a guest basis."[16]

These noncommercial organizations are candid about wanting to make the Internet a place of noncommercial, public service access for historical information, and have openly challenged the commercialism of the CD-ROM publishers:

> The real reason RootsWeb got started was the feeling that the data being sold to genealogists on CD-ROMs was grossly overpriced and, even then, not nearly enough new data was being made available to the community each year. . . . Our long-term goal is to be able to subsidize the development of significant new resources for the genealogical community.[17]

There are thousands of individual home pages associated with genealogy research. For example, a promotional description for one person's page reads, "Explore the Frontier of Time and Space—Genealogy is enhanced by being placed in historical context; History is illuminated by individual experience. My site uses family memoirs and research. . . . For fun, my other passions, *Star Trek* and the Seattle International Film Festival, have pages [too]."[18] On these pages, which are sometimes absolutely fascinating in their attempts at personal stylistic expression, people can create personalized multimedia projects centered around their hobbies and interests. The majority of them include family history charts in one form or another (many of them available for downloading in a GEDCOM format), while many include family photographs (both contemporary and antique), archival documents, family stories, pictures of family homes, and links to favorite genealogy research sites. On the whole, such Web pages can be characterized as personalized, down-home, and highly individualized home pages built around personal family histories and interests. There are also family surname pages, usually coordinated by one individual, which might be associated with either a surname-oriented family society or an e-mail–based discussion list, and which feature the collaborative efforts of people who share a research and/or kinship connection with a family surname.[19]

The Discourses of Computer-Mediated Genealogy

As with any ethnographic project, it is quite difficult to contain and impose order on the knowledge I have gained about how and why people engage in this practice of genealogical research, but I can identify at least four motives that either draw people into genealogical research or keep them there once they are involved. Ethnographic correspondence and other sources have provided detailed personal narratives about the way people make use of their CD-ROM and computer-mediated resource networks, commercial and non-commercial—in conjunction with more traditional genealogical resources—to accomplish their desired goals.

The first and most overtly expressed motivating factor in genealogical research is related to the *construction of personal and cultural identities* in our postmodern world. For many people, knowing about their ancestral heritage helps them situate themselves. Who are my people? Where do they come from? Who *am* I? What is my cultural heritage? This knowledge contributes to a sense of class, racial, social, and ethnic identity, as well as a feeling of connection to a flow of history as lived through the kinship-connectedness of generations of social actors and players.

Some of my correspondents spoke of a spiritual or deeply emotional connection to the past as a key element of their sense of identity. For example, one researcher said,

> Sometimes to better know ourselves, and where we are going, we need to know where we come from. A sense of identity and "who am I" are critical to our well-being. Sometimes knowing the hardships our ancestors overcame helps us overcome hardships and difficulties in our own lives because your very existence is testimony to the fact that survival is possible.[20]

As may be evident, there is a deeply affective aspect to family history research that becomes almost tangible when one browses through the personal testimonials and individual home pages of amateur genealogists or reads a few conversations on a discussion list or two. The needs that drive people to purchase software or to participate in discussion groups range widely from the purely pragmatic exchange of information ("Does anyone know who the parents of William Arledge, died 1724, might have been?") and mechanics of dealing with the software ("Could someone help me figure out how to translate my GEDCOM file into FTM format?") to the more private and personal exchanges marked by a sense of familiarity and kinship, in which people express their deeper longings for connectedness and identity. Some of them find this connectedness with the past as they discover names and knowledge about their ancestral families; others find it through their connections with newly found "cousins" (often many, many times removed, genetically, in terms of American kinship calculation) and with other kindred spirits who share their passion and enthusiasm, genetically related or not. As one amateur historian articulated,

> I've wondered why I'm so driven by genealogy. It started a long time ago, I think. I wondered why I was the way I am. Then why dad and mom are the way they are. And then my grandparents. And so on. It starts with a desire to understand one's own self, I think. And our ancestry *is* who we are, isn't it? . . . I recently had to make some major changes in how I live my life. But I have a goal and a desire to make more out of my life . . . to make it *mean* something . . . thanks in large part to my involvement in genealogy.[21]

The search for interpersonal and familial connectedness is a major factor that motivates people to participate in genealogical research.[22] The possibility of creating new "family" links is enormous. As researchers find others who share common ancestors, discourses of "cousinhood" abound: some serious, others tongue-in-cheek. At what point in American society is it no longer appropriate to consider a person with some small percentage of shared "blood" a cousin? The new genealogical research raises this problem, since researchers often connect with others who share a common ancestor, albeit several hundred years or more in the past. How does one categorize these people who are not "family" in traditional terms but do, in fact, technically share kinship ties? What does it mean to be in community with a group of people, discovered through CD-ROMs, e-mail discussion groups, and Web pages, who share not only some of your genetic material but also (sometimes) your personal interests, your culture, or your heritage (or who do not)? These are deeply personal affective questions with which many researchers grapple. In the meantime, the accepted term in the online genealogy community is just "cousin," until a better kinship category comes along.

The second motive for engaging in computer-mediated genealogical research might be called *epistemophilia,* or the quest for knowledge. This differs from the first (the quest for personal or cultural identity) in that it is a more abstract, intellectual process.

> This "hunger" for research began in 1964 when I was in my teens. My maternal grandparents ran a big old Victorian rooming house just down the street from Washburn University in Topeka, Kansas. I was exploring the attic one gloomy and cold Saturday morning and found a shoe-box over-flowing with old photographs. The eyes of the people in them just intrigued me. I thought that this treasure may have belonged to the old man the house belonged to before my grandparents bought it; so I gathered up the dusty things and went down stairs to the kitchen where I could always find my Grandma Kerber. I showed her my "find" and asked if I could have them. She said, "Well, yes—if you take care of them." I asked her who they were and she told me family.
>
> My parents divorced when I was five and my Dad brought me up, so finding these people from whom I came was very exciting to me. I went out that day and bought my first album which has grown to two file cabinets, genealogy association memberships, a computer and 33 years of haunting cemeteries and sitting in dark libraries doing research on warm, sunny week-ends. Why? I've asked myself many times. There are so many reasons—but it all boils down to a "need" somewhere to "know."[23]

Many amateur researchers find that the process of genealogical research awakens—or satisfies—an intellectual hunger for historical knowledge, and leads to a better understanding of larger social and cultural history. In this way, genealogy often opens research avenues to the nonacademic investigator that have in general been reserved for the academy. As one of my correspondents stated, when asked why they do it, "The puzzle-solving aspect is certainly one strong component, but I also find that the process of doing *research* is very satisfying. I imagine scientists, historians, archeologists, etc. have the same satisfaction in the hands-on aspects of their fields."[24] I have found that many of the people who engage in this sort of historical research are people who, by nature, are intellectually curious (even if they are not professionally trained in a research field) and who find the challenges of the research process itself to be stimulating and satisfying in a profound way. This is closely related to the gaming or puzzle-solving aspect described below, but with an added intellectual component that goes beyond just seeking their personal histories.

> I haven't really touched on the feeling of "family history" that genealogy invokes, but that is certainly a part of the puzzle. Before my interest in genealogy began, I had no concept of working life and conditions in 1850s Scotland. Now, I have that appreciation. And I think that the ability to put known relatives into a place in general history has a great value as well. (I won't quote that old Santayana phrase "Those who forget history . . .") Enough said! There's obviously a hobby value to genealogy as well. I'd like to think that when I retire I'll be able to pursue genealogy (and travel related to genealogy) into my old age.[25]

Many people use this type of personalized research as a springboard to understand the larger and more abstract patterns of historical happenings and the various cultures, politics, and social movements that accompany them.

The third motivating factor, one that may not be the initial instigating factor but that is often foremost in *keeping* people involved, might be called *the gaming factor*. Talk of the addictiveness of the genealogy "habit" and one's compulsive behavior as a practitioner is frequent in the genealogical community. One person said, "Researching genealogy has become that well-known obsession that those of you who love genealogy are probably afflicted with. My goal is to not only find information about my family history but to allow others to broaden their searches too."[26] Another responded,

> The part about it that seems to be so addicting is that for every answer found, there are a lot more questions raised. Who were his or her parents, brothers, sisters, etc. Why did he/she move to this place? I guess it is a puzzle that never ends. We add pieces together, and occasionally a larger section to ours, only to find out the puzzle is larger than we anticipated. (I like doing puzzles.)[27]

This whole endeavor of computer genealogy is like a huge game in which information about people, births, deaths, marriages, family relationships is exchanged in a giant economy. These fragments of genealogical knowledge become the currency around which all interactions take place. When I first started as a participant-observer, for example, I was necessarily in a major information-gathering mode. Then as I built up a huge database of information and began to ease up on the intensity of my own data-gathering quest, I became the sought-after rather than the seeker. Once a researcher puts her names into the networks, on the Web pages and lists, there is a deluge of e-mail from people wanting information *from* her about what's in her database. And it's part of the ethics of this "game" to share generously, since one expects the same from everyone else.

What is the goal of the game? To fill in all the blanks in one's ancestry. It's impossible, of course, since the blanks keep going back, and back, and back . . . eternally. For many, it becomes an obsessive/compulsive need to find more clues to help fill in yet one more blank. And the resources are out there—the CD-ROMs and other resources become the locus for a gigantic treasure hunt, and the fragments and clues are available through archival research and/or other people who may have already solved that piece of your puzzle for you.

A side benefit of all this exchanging of fragments of genealogical information is that, since one must correspond with people in a sociable way, there's a great deal of social grace involved, and other researchers are, on the whole, extremely friendly, helpful, and generous. It's an amateur game, and except for the commercialization of the software and a few Internet sites, the whole economy of knowledge exchange among participants is relatively free of commercialization and business interests. Since this is also a very collaborative endeavor (working on constructing larger puzzles than just one's immediate family line, for example—like all the descendants of someone with a certain surname), there is a tremendous amount of community building and collaboration. In this way, "playing genealogy" becomes in many ways comparable to other CD-ROM games—a potentially infinite puzzle to be pieced together, addictive and at times overwhelming in its potentialities. It's at the same time both an individual and a collaborative puzzle-solving effort.

The fourth motivating factor, then, is the social—the *forging of communities* and interpersonal connections that occurs through correspondence both online (e-mail, family chat rooms, discussion groups, mailing lists, Usenet groups) and in traditional ways (telephone contacts, letters). There is a "genealogy community" out there connected by the Internet and e-mail and other interpersonal connections, sharing a bond of common research interests that is acknowledged by the commercial and noncommercial groups alike. Geocities, a very popular Web site home for genealogists and others (it provides free space for personal home pages), organizes its Internet space and its users around a geographical, spatial metaphor of theme-based communities and neighborhoods.[28] The genealogy community, though an abstract collective of folks from around the world, openly considers itself a community—but, importantly, a collectivity of *named* individuals rather than anonymous members or people associated with family names. Individuality and the power of individual initiative are strong cultural values in this computer-mediated culture that celebrates the historical continuity of the collective bond (i.e., family). Yet the power of the (noncapitalist) collective is also strongly celebrated. For example, on the noncommercial RootsWeb home page is a page entitled "Who's building RootsWeb?":

> The simple and mostly wrong answer is Karen Isaacson and Brian Leveric. We've assembled the hardware, Net connections, and software. The better answer is the data providers and the online genealogy volunteers, because everything that happens at RootsWeb is the result of their efforts. The best answer is the whole genealogical community, because in the ultimate sense RootsWeb is like all other genealogical resources—it's the result of all our collective efforts.[29]

Individuals are important in the genealogy community, and are often acknowledged and become known on a first-name basis. Kinship groups—those researching the same family surnames—and geographically based groups are also important. But there is also a sense of, and a respect for, a larger collective, that all-encompassing community (an admittedly problematic concept, but one that is mythologized), which occupies a public space and is entitled to the documents and the heritage of the mythicized public sphere.

Genealogy and Cultural Studies: Postmodern Kinship in a Virtual World

Academic ethnographic studies of online communities have become a new and valuable trend in cultural studies in recent years, and provide rich data on the mechanisms of how, and around what types of issues, postmodern communities form and flourish. On the other hand, the few academic studies of new media technologies such as CD-ROMs have most often focused on textual or technological characteristics rather than the way such technologies have been appropriated and used by people in everyday settings. Scholars in general have shied away from examining the historiographical implications of the genealogical endeavor, perhaps because it has been perceived for so long as an aristocratic hobby that only served to reinforce class and race privilege. However, genealogy—as a practice, a process, and a social artifact—lends itself to the analytical scrutiny of cultural studies, since it raises boundary-blurring issues related to postmodern notions of cultural and social identity, historiography, geography, and cultural space.

One of the many insights I have gained in this study about the ways new technologies have been adapted to ancient practices is that the new research technologies have transformed and reconstituted those practices (and their very meaning) to the people who participate in them. The sociologist Anthony Giddens has characterized (post)modern social life "by profound processes of the reorganization of time and space, coupled to the expansion of disembedding mechanisms—mechanisms which prise social relations free from the hold of specific locales, recombining them across wide time-space distances" (Giddens 1991:4). This recombination has nowhere been as clear to me as in the creation of bonds of commonality across time (familial links over centuries) and across space (globally dispersed and electronically linked kinship groups) that I have found in the computer-based genealogy process. Giddens continues with an insight particularly appropriate to the consideration of history in general, and genealogy (personalized history) in particular:

> The media, printed and electronic, obviously play a central role.... Mediated experience... has long influenced both self-identity and the basic organization of social relations. With the development of mass communication, particularly electronic communication, the interpenetration of self-development and social systems, up to and including global systems, becomes ever more pronounced. The "world" in which we now live is in some profound respects thus quite distinct from that inhabited by human beings in previous periods of history. It is in many ways a single world, having a unitary framework of experience... yet at the same time one which creates new forms of fragmentation and dispersal. (1991: 4–5)

In this social order, Giddens argues, "against the backdrop of new forms of mediated experience, self-identity becomes a reflexively organized endeavor." To Giddens, the "reflexive project of the self" consists of the construction of "coherent, yet continuously revised, biographical narratives" (1991: 5).

Ironically, it is this postmodern condition that has impelled so many people to seek to understand previous periods of history, and they are using the most effective technologies that postmodernism has to offer (or, more cynically, that postmodern corporations have to sell) in their quests to "know" the ancient worlds of their ancestors. In a postmodern world, too, "kinship" becomes redefined. Although the concept of the family unit has been shrunk in the late twentieth century to the notion of the nuclear family (and the ideal of the mythical nuclear family itself has been exposed by the media as dysfunctional and fragmented by divorce or single-parent families), the phenomenon of online genealogical connections demonstrates a radically new concept of kinship, still based on some semblance of genetic connections but also deeply invested with ideals of friendship and collegiality. History, then, for amateur historians engaged in genealogy, blends the personal and social, becoming both private and public, bridging both geographical space and historical space. The ends they aspire to are multiple: the knowledge of the past, the knowledge of the self (and sense of identity) that this may provide, and the intimacy of kinship and friendship forged through community with like-minded "kin" and research colleagues. As these amateur historians find others engaged in the same pursuit, they form electronic communities of seekers.

NOTES

1. A major example is the work of the anthropologist David Schneider, whose 1977 classic on American kinship revolutionized the way anthropologists and sociologists conceptualized "family" in American society.

2. A major institutional force has been the New England Genealogical and Historical Society.

3. See article by Shelly Reese, as well as numerous Web sites, such as http://members.aol.com/Scabbegen/index.htm, and the soc.genealogy.african Usenet news group. Also see recent practical guidebooks for African American genealogical research such as Blockson and Fry (1997), Beasley (1997), and Johnson and Cooper (1996).

4. Mary C. Waters (1990) discusses issues of choice regarding ethnicity and ancestry in identity formation in America.

5. Virginia Parmenter, personal e-mail correspondence, 4 June 1997.

6. In addition to their extensive Salt Lake City library holdings catalog, the star attractions of the LDS Family Search System are the Ancestral File CD-

ROMs (linked collections of genealogical pedigrees) and the International Genealogical Index, or IGI (a compilation of culled vital public records from the early 1500s to 1875).

7. Virginia Parmenter, personal e-mail correspondence, 4 June 1997.

8. Virginia Parmenter, personal e-mail correspondence, 4 June 1997.

9. Barbara L. Hill, personal e-mail correspondence, 3 June 1997.

10. "SPATCH," personal e-mail correspondence, 6 June 1997.

11. For example, see Stevenson (1969).

12. http://www.familytreemaker.com/.

13. http://www.usgenweb.com.

14. http://www.dsenter.com/faqs/ushistry.htm#abou, 30 May 1997.

15. http://www.rootsweb.com.

16. http://www.rootsweb.com/, 10 October 1997.

17. http://www.rootsweb.com/rootsweb/more-about-rootsweb/missions.html, 10 October 1997. Among other noncommercial online organizations, the monthly *Journal of Online Genealogy* (http://www.online genealogy.com/), started in 1996, is self-described as "a free e-zine which focuses on the use of online resources and techniques in genealogy and family history." In addition to how-to articles, it includes press releases from genealogical companies and organizations, and a feature called "Online Success Stories," which "gives examples to researchers of how others have successfully used online resources to achieve research results." The *JOG* is associated with Matthew Helms's Toolbox Internet Marketing Services, Inc., which also hosts Helms's Genealogy Toolbox, a well-known centralized source of genealogical information on the Web.

The most publicized Web site of links to all things genealogical, however, and often highlighted in magazine features on the amateur genealogy boom, is Cyndi's List of Genealogy Sites on the Internet, by Cyndi Howe (http://www.oz.net/~cyndihow/sites.htm). This noncommercial site, coordinated, like many, by a single individual as a labor of love for fellow genealogical researchers, includes ethnic/cultural categories of research links, such as African American, Eastern Europe, Germans from Russia, Hispanic, Huguenot, Jewish, Mennonite, Native American, Poland, Quaker, Switzerland, as well as sources of information related to adoption, books/microfilm, cemetery/obituaries, census-related sites, heraldry, how-to, land records, libraries/archives, medieval, military resources, photographs, royalty, ships/passenger lists, and even recipes and family traditions.

18. Description/summary of Web site by Marjorie Jodoin, http://www.geocities.com/Heartland/Meadows/5209/, June 1997.

19. An example is the Arledge/Aldridge Family Homepage (http://www.tx3.com/~arledge), which is the home page and repository site for a team of fifty to seventy-five researchers who are connected through their

interest in the Arledge family and its variants (Aldridge, Allred, Aldrich, Aldred). It was established and has been maintained through the labors of one individual, Clay Fulcher, who contributes a great deal of time and energy (and funding to rent the Web space) because of his personal dedication to the group and its purposes. The Web site has become a centralized location for all the group members to electronically "gather" and publish their family trees and other information.

20. Rebecca Ellis, 10 June 1997, Medieval Genealogy Discussion List (Gen-Medieval).

21. Clay Fulcher, personal e-mail correspondence, 1 June 1997.

22. For some, the search for personal and family identity has been thwarted because of adoption, and new information technologies provide new opportunities for adoptees to find their "birth family," as in this personal narrative:

> In 1911, my father was born out of wedlock in Detroit, Michigan and given to a couple who took him to live with them in Toledo, Ohio. . . . He grew to manhood and married and a time came when he needed a birth certificate and adoption papers in order to work for the federal government. He was unsuccessful in finding the needed documentation and he turned to his birth mother for help. It was soon apparent to his birth mother that my father had never been [legally] adopted. She sent him his birth certificate. My father searched all of his life for the man who was shown on the birth certificate as his father. He did not have the resources that we now have and died never knowing anything other than the name of his father.
>
> A few years ago I decided to pick up where he left off and try to find the family. I was successful in finding his birth father on the census records and have been able to trace the surname of his father back to 1850 in New York state. I discovered GENWEB and was poking around in the Genesee county GENWEB site and saw the name of Linda Dent who would do look-ups in the Flint Library. I sent her an e-mail and asked her if she could check the city directories for the name of my father's birth father. I knew he had married and lived in or near Flint. I wondered how long he lived and if he had any half siblings. This lovely lady went not just the proverbial extra mile but an extra hundred miles. Before she was done I knew that my grandfather never had any other children other than my father, had the obituaries of almost everyone in the family and have since been in contact with a second cousin. It doesn't stop there. With the information that Linda sent I have been able to take those pieces and have traced my great-grandmother's family back to the Revolutionary War. I now have the pension file of my

> 5th great-grandfather, know all the names of his children and have found records on other family members.

Glydie Ann Nelson, Anchorage, Alaska, printed in *Journal of Online Genealogy* as "USGenweb Site Aids Researcher," March 1997 (http://www.onlinegenealogy.com/success/suc0181.htm).

23. "SPATCH," personal e-mail correspondence, 6 June 1997.

24. Barbara L. Hill, personal e-mail correspondence, 3 June 1997.

25. John Brebner, personal e-mail correspondence, 5 June 1997.

26. Felicia Hazelton, http://www.earthlink.net/~hazefam/, June 1997.

27. Jerry Donovan, personal e-mail correspondence, 4 June 1997. However, the negative side of the addiction is also invoked, such as this case:

> I worked feverishly for about a year on genealogy. I got hooked. It became an obsession. I even spent a whole week in the LDS Genealogy Library. Made the trip to Utah just to do that. Well, if I wanted any life at all, I had to quit cold turkey. I'm very reluctant to get back into it. It really fascinates me, but it's almost like a gambling habit only I lose time instead of money. It's more beneficial than gambling, but it can make my whole life go to pot!

Mary Hipp, personal e-mail correspondence, 11 January 1998.

28. http://www.geocities.com/.

29. http://www.rootsweb.com/rootsweb/more-about-rootsweb/founders.html, 10 October 1997.

WORKS CITED

Beasley, Donna. 1997. *Family Pride: The Complete Guide to Tracing African-American Genealogy*. Indianapolis: Macmillan.

Blockson, Charles L., and Ron Fry. 1997. *Black Genealogy*. Baltimore: Black Classic Press.

Fulkerson, Jennifer. 1995. Climbing the Family Tree. *American Demographics* 17, no. 12 (December): 42–50.

Giddens, Anthony. 1991. *Modernity and Self-Identity: Self and Society in the Late Modern Age*. Stanford: Stanford University Press.

Johnson, Anne E., and Adam Merton Cooper. 1996. *A Student's Guide to African American Genealogy*. Phoenix: Oryx Press.

Reese, Shelly. 1995. Shrouded by Slavery. *American Demographics* 17, no. 12 (December): 42–50.

Schneider, David. 1977. *American Kinship: A Cultural Account*. Chicago: University of Chicago Press.

Stevenson, Noel C. 1969. *Genealogical Evidence: A Guide to the Standard of Proof Relating to Pedigrees, Ancestry, Heirship and Family*. Laguna Hills, CA: Aegean Park Press.

Wagner, Anthony Richard. 1960. *English Genealogy*. London: Clarendon.

Ward, Geoffrey C. 1995. Family. *American Heritage* 46, no. 4 (July–August): 16–17.

Waters, Mary C. 1990. *Ethnic Options: Choosing Identities in America*. Berkeley: University of California Press.

Chapter Nine

Fantasies of Mastery or Masteries of Fantasy?

Playing with CD-ROMs in the 5th Dimension

Vanessa Gack

This essay explores children's positions of power as constructed in relationship to two CD-ROM games in the context of an after-school program called the 5th Dimension (5th D). I look at the ways two kids create and occupy alternative spaces in the 5th Dimension and argue that these alternatives provide them opportunities to occupy positions of power. Freddi, a twelve-year-old girl disdained by her peers, and Adam, a ten-year-old boy stigmatized by the label of attention deficit hyperactivity disorder (ADHD), use their play with CD-ROMs in the 5th D as means to insert themselves into fantasies of mastery. I will show that children's participatory culture is not totally dependent on "real life" but is also a function of the imagined; thus CD-ROMs and the 5th D offer important opportunities for children to inhabit vastly different roles and express vastly different forms of competence.

I argue that Freddi's and Adam's positions of mastery in fantasy-mode differ from those that they appear to occupy outside the 5th D. Mastery in this sense refers to the different ways Freddi and Adam entered into relations of power—over the game worlds, their peers, and supervisory adults. Drawing from their practice of mastery over game worlds and others, I treat their play as performances of fantasy and power. Where they are successful, I employ the term "masteries of fantasy and power" to refer to their mastery over three domains: self, social world, and future worlds.

Field Site: The 5th D

Located at Santa Fe Beach in a Boys and Girls Club, the 5th D is held in the afternoon during the school year. The Boys and Girls Club provides a room where computers and software are accessible; it also provides the salary for a "site coordinator" (staff person) who supervises the after-school program independently of the rest of the club offerings. The University of California San Diego (UCSD), in turn, supplies undergraduates who are enrolled in field methods classes. These undergraduates fulfill class fieldwork requirements by attending the 5th D twice a week. They pair up with kids, play computer games, solve problems, and write notes about their interactions. These field notes provide researchers like me with a kind of record of the kids and their interactions—with the undergraduates, computer games, and other kids.[1]

An important factor in all the interactions around the computers is the flexible structure of the Boys and Girls Club and the 5th D. Kids come to the club for many reasons: because their parents work, to hang out with friends, to play in a semi-supervised environment. These kids spend their time participating in a variety of sport, art, and free-play activities; participation in the 5th D is voluntary.

Located at the south end of the club and separated by a glass wall and wooden doorway, the 5th D competes with the other activities at the club for the attention of the kids. In the 5th D room stands a decorated three-by-five-by-one-foot wooden maze used by participants to symbolically enter and move around in the world of the 5th D. UCSD students and 5th D kids make tokens or "avatars" from Polaroid head shots and place these representations of themselves in the maze, a 3-D representation of the 5th D and its activities. 5th D "citizens" travel through the maze by completing some or all of the tasks in each room of the 3-D maze. They choose activities from among the rooms (usually computer games, but sometimes other activities such as exploring the WWW, producing video, or making origami figures) and keep track of their progress in personal folders. The activities are a mix of play and education. Each activity has a task card with different levels of task completion (beginner, good, and expert) and different consequences for each level (usually the higher the level completed, the more latitude one has in choosing the next room of play).

This world, complete with maps, tasks, and rules, is governed by a mythical, transgendered entity called the wizard. The wizard reads letters that kids write when writing is specified on task cards or when a child wants to tell the wizard about what is going on in his/her life in the 5th D. The wizard serves as an arbitrator of this rule structure as well as for the disputes between the kids, undergraduates, and technology. When games fail or people disagree, it is convenient to complain to an off-site entity, in this case a wizard who appears to the citizens of the 5th D through electronic mail and takes responsibility for solving problems and fixing machines.

Driving this organization is a reward structure for the participants. Citizens who complete task levels (which may include writing to the wizard, recording game-playing strategies as hints for other 5th D citizens, sending letters to 5th Dimensions located elsewhere, and other kinds of literacy activities) receive encouragement, free passes to popular rooms in the maze, and credit toward becoming a "young wizard's assistant" (YWA). YWAs are citizens who have completed ten games at the good level and ten games at the expert level as well as other requirements set forth by the wizard, including an application essay. YWAs, in turn, receive special privileges from the wizard, including free play of games inside and outside the maze. YWAs have their own set of computer games that are outside the maze, which include *Sim City 2000* and *Sim Town*; they are also allowed to suggest and test out new computer games brought to the 5th D. In addition, they are encouraged to write task cards and share computer and 5th D expertise with their fellow citizens.

Kids in the 5th D

I have chosen to focus on two kids known at the 5th D for their visible engagement with high-tech CD-ROM and simulation games. Both are acknowledged by peers and adults alike as "experts," yet both are, in general, disliked. Freddi, who appears pretty and smart to adults, is nonetheless unpopular at the Boys and Girls Club. Kids call her "dog face" and "Freddi the dog," among other jeers and insults. Freddi takes refuge from her peers by playing games and appealing to adult authority, rules, and intervention within the 5th D. Since play is relatively unsupervised and unstructured outside the 5th D, the 5th D serves as a kind of haven for both Freddi and the other child, Adam. Like Freddi, Adam goes to the 5th D to assert himself. Although Freddi is successful in school, Adam has been held back. Stigmatized for having ADHD and teased about being overweight, Adam seeks refuge from the rough activity he encounters among boys outside the 5th D. His alternative is to immerse himself in "fantasies" of technological control and power. It seems that the unsupervised behavior and free flow of cruel language in the rest of the club help to reinforce Freddi's and Adam's involved game play.

Adam and Freddi are both marginalized at the Boys and Girls Club. Freddi, while hardly "stigmatized," is perhaps more marginalized than Adam. Despite the label of ADHD that has been given to Adam, kids at the 5th D still respect and acknowledge his skill with *Sim* games, whereas the kids make fun of Freddi even when she is expertly building game worlds. Adam seems to take solace in his solitary involvement with *Sim City 2000* and Freddi with the undergraduates and 5th D procedures that shield her. Each child has a reputation for employing different kinds of expertise: Adam interacts with those who seek him out for his *Sim* expertise and Freddi gets deeply involved in the micro worlds of the games. Both Adam and Freddi are known as experts at manipulating 5th D rules to gain access to the simulation games available only to YWAs. In this context of achievement with CD-ROMs in the 5th D they are both successful.

CD Media

Recent work on CD-ROMs situates them in a tradition of prior media, particularly TV, film, and animation. Play with "dramatic" media and CD-ROMs can be framed in similar ways. In *Textual Poachers: Television Fans and Participatory Culture* (1992), Jenkins explores the relationships between communities of television fans and their texts. Jenkins sees the "borderlands between mass culture and everyday life" (3) in the interpretive practices of fan communities. He sees how fans use, reuse, and transform program materials into resources that perform multiple functions in their lives—in ways that resemble Adam's and Freddi's play with CD-ROMs. Like Jenkins's and Tulloch's communities of fans, the children that I study develop unique relationships with CD-ROMs within complex social networks that mediate their practices/constructions of meaning.

The 5th D differs from Jenkins's and Tulloch's fan communities in that it includes external authority figures (adults) to police the community. Whereas fans tend to police themselves, the 5th D is regulated by authority figures such as the wizard, and often by the undergraduates. As is evident from the undergraduates' field notes, adults play key roles in framing children's behavior. This is important for understanding children's' cultures. Kids are put in the position of making cultural choices while at the same time being both supervised and judged by adults; these judgments affect children in ways that have real consequences in their everyday lives.

Another difference between research on TV and research on computer games has to do with cultural notions of value inscribed on to audiences and their behavior. Television fan behavior is traditionally accorded a low status, whereas interacting with computers has the potential for high status appeal. Although the use of computers is not considered a lower-class consumption/taste, the metaphors of addiction still come up. In *Moving Images*, Buckingham interviewed parents about their attitudes toward TV and found that many parents use the metaphor of addiction to describe their children's viewing habits. In research at the 5th D I found similar fears among parents (especially Adam's mother) about their children's use of computers. I also found, however, that parents' fears are often mediated by another, often stronger, parental concern about knowledge/power. In many cases (as with Freddi's parents) I found that the cultural capital (Bourdieu 1984) associated with computers and computer expertise can outweigh their concerns about addiction, particularly in the case of CD-ROMs (as opposed to arcade games).

Children's play with CD-ROMs can be seen to blur the boundary between fantasy and reality. As Walkerdine (1990) suggests, children's insertion into practices is not totally dependent on "real life," but is "related to the imagined through their insertion as subjects within storytelling, the media and other cultural practices" (142). Nonetheless, the relative importance of realistic elements versus children's imaginative contributions is unclear. Tulloch argues that for television audiences, "perceptions of the show's degree of realism" serve an important function. For Tulloch's fans, pleasure and recognition of relevance were deeply intertwined (Tulloch and Jenkins 1995, 109). Tulloch's fans judged the degree of a program's relevance not only in terms of its place along a continuum of fact and fantasy, but also in terms of its relevance to a particular interest, community, or communities. Tulloch found, for example, that teenage girls described relevance in terms of "real life"—"real personal relationships" (as in the soap opera genre); boys and men in terms of relations to other science fiction texts (action, graphics, explosions). Tulloch's broad definition of realism—in terms of degrees of relevance—is of particular interest to our study of Adam's and Freddi's imaginative practices with CD-ROMs in the 5th D.

Mastery over Games and Game Worlds: Sim City 2000 *and* Sim Town

The ways Freddi and Adam represent themselves in two CD-ROM simulation worlds, *Sim City 2000* and *Sim Town*, suggest that they come to the 5th Dimension because it affords them a kind of control over their labor and creativity that they are not getting at home or at school. While each is productive in the context of the 5th D, both are positioned as "problem kids" by the people around them. I treat their play as performances made visible in their interaction with CD media. I am interested in these performances not as isolated events, nor as problems, but in terms of their relationship to the real-life context in which they are produced. By context, I mean not only what is going on in the computer games, but what is going on around the games and club, as well as in the kids' lives. My aim in this section on mastery over game worlds is to show the traffic across boundaries marking inside/outside simulated worlds and fantasy/reality.

Freddi and Adam create game worlds that reflect their worlds beyond CD-ROMs. Issues of gender, class, and family arise in relation to Adam's play with *Sim City 2000* and Freddi's play with *Sim Town*. I use transcripts from videotape segments and excerpts from field notes to examine different forms of play that appear to achieve similar goals.

Sim City 2000 is the next generation of *Sim City*, a cult favorite once sold on floppy disks but copied and shared among friends. While it seems to have its roots in LEGOs, Erector Sets, and simulation games (such as those used in the military and in ecological modeling), *Sim City* was a programming feat—not only for its ability to model relationships between complex variables in an aesthetically pleasing way, but by virtue of memory requirements that were small enough to fit on a home computer. *Sim City* was so different from most of its predecessors that it has been called its own genre and has spawned dozens of other simulation games, including *Sim City 2000* and *Sim Town*. With *Sim City 2000* and *Sim Town*, both available on CD-ROM, the MAXIS corporation produced more sophisticated interfaces with potential access to more parameters and complex modeling. These "new" *Sim* formats gave MAXIS the potential for a wider consumer base and a higher percentage of consumers willing to pay CD-ROM prices.

Sim City 2000 is a simulation game with a sophisticated graphical interface that allows users to design and build virtual cities. With a complicated tool bar similar to standard imaging applications such as Adobe Photoshop, *Sim City 2000* has a relatively high learning curve (in that it requires the user to invest a fair amount of time to gain expertise). The tool bar introduces concepts of zoning—industrial (tan)/commercial (blue)/residential (green)—and monitors the growth of those zones with small bar graphs, indicating which zones a city needs for further growth. Other tools afford the user choices between kinds of buildings and infrastructure such as transportation, water, and power. Other icons allow the user to view the city from multiple vantage points (rotated, near, and far), change terrain, and control various pop-up screens that allow the user to monitor details of the city's progress.

Explicit in the game's structure is the goal of urban growth in an economy of financial and industrial planning. New players choose between easy (20k), medium (10k), and hard (10k bond) levels, then set a budget that provides for the city's growth, income, and expenditures. This budget can be adjusted at any time, but if the city runs out of money, then transportation, the economy, and new production fail. The city may appear doomed, that is, unless the player has saved an earlier, more stable version of the city or knows the secret code to receive unlimited funding.

Sim City 2000 affords players several options for "subverting" its financial planning and industrial growth goal structures. The "load saved city" function allows players to recover financially unstable cities as well as those plagued by disasters. Embedded in the task structure of the game is the option of turning disasters (fires, floods, tornadoes, plane crashes) on and off, thereby allowing kids the thrill of "knocking over their blocks" safe in the knowledge that their cities can be instantly recovered. Moreover, a "secret" back door command programmed by the game's designers allows players who know the "code" to seek unlimited funding.

It is important to note that access to both these games was given to children at the 5th D who had achieved competency in twenty computer games in the maze. These gifts to the young wizard's assistants were intended for play by YWAs or children with free passes. There is no task card for these games. The content is less overtly "educational." Expertise is often attained through longer and more sustained interactions with the game and/or through other experts in and outside the 5th D (the MAXIS Web site, other *Sim City* sites, YWAs, undergraduate experts, etc.).

Adam seemed obsessed with *Sim City 2000* and all the technology that makes it possible. From a field note written by J.S. (4/18/95), we learn of Adam's fascination with technology. J.S. described Adam's interest in an old broken Apple II computer: "Adam asked me if we could take it apart so that he could take the motherboard and power supply home." Adam told J.S. that he wanted to play "Mini-city" with them, using the computer chips as the buildings and metallic connections as the streets. In trying to understand how computers were put together, Adam imagined the parts as themselves composing a kind of *Sim City*.

I see this image of Adam trying to build a simulation city out of computer parts as a kind of symbol for Adam's burgeoning mastery. While I cannot claim access to Adam's subjective world, I do see the effects of that world on the space around him. In this instance, Adam was learning how to transform an imaginative vision of *Sim City* into a real city made out of computer parts. I see Adam's imaginative play with computer parts as a kind of "potential space." According to Winnicott's perspective on potential spaces, fantasy (subjective experience) and reality (objective experience) can become intermingled in an act, perception, or physical object (Deri 1978, 46). Observing a small boy drawing a picture, Winnicott sees "the pleasure that self-discovery gives [the boy] and the possibilities of what he may draw" (Davis and Wallbridge 1981, 167). Adam, however, appeared to derive more than pleasure from his play with computer parts. Adam explicitly used his knowledge of computers and his mastery of *Sim* as a license to transgress rules. By asking to take an item that belonged in the 5th D, he was asking to break the cardinal 5th D rule. I argue that this effort to take technology home represents a kind of boundary crossing for Adam, who subsequently developed a reputation for this.[2] In transforming (what other people called) trash, Adam was learning creativity of the sort employed by scientists, writers, movie makers, computer programmers, policy makers, and so forth. Moreover, he was trying to take his expertise outside *Sim City* and outside the 5th D.

As Adam's skill with *Sim City 2000* grew, so did his fascination. He became especially excited when the game informed him that the population of his city had reached a new classification: metropolis. He appeared more challenged by problem cities, such as those that had a low public opinion rating. Perhaps motivated by his own issues of self-esteem, Adam grappled with these problem cities, methodically rebuilding them, step by step. A low rating demanded that Adam use his expertise to decide what elements a city was lacking. Was safety an issue, or taxes? Adam explained that if safety was a concern of the people, then more police were required. He explained his rationale to undergraduates, who wrote about his progress in their field notes. One undergraduate, for example, described how Adam's fascination with *Sim City* escalated. C.M. (2/6/95) wrote that Adam's biggest accomplishment of the day was when the city's population gave him a 100 percent approval rating and Adam achieved a city score of 977 (out of 1,000). C.M. wrote that when Adam's mother picked him up, Adam left excited about getting to megatropolis next time.

Adam seemed to derive certain pleasures from his mastery of *Sim City 2000*. He located himself outside the worlds he created—and enjoyed creating his worlds as well as destroying them. Research on computer gaming suggests that part of the appeal of these fantasy games is the way games can be named, saved, and recovered (Ito 1997). Adam could, for example, ruthlessly destroy his city before an audience of onlookers, safe in the knowledge that his city could be restored in its saved form at any time. Not only did he get to feel the power of creating and demolishing cities with a click of a mouse or keystroke, but he got an opportunity to display his expertise.

Freddi displayed a different kind of mastery in her play with *Sim*

Town. *Sim Town* was also one of the higher-end computer games available at the 5th Dimension. Like *Sim City 2000*, it was available only to children who have become YWAs or who have free passes. But whereas *Sim City 2000* provides an omnipotent, bird's-eye view of the city, in which inhabitants are represented as specks on the screen, *Sim Town* emphasizes inhabitants and their relationships within a community.

Within this space Freddi created a computerized doll house, complete with Freddi, Mom, Dad, and Sister dolls, set in a doll neighborhood. Since Freddi located herself in the scenarios of the game, I explore Freddi's play as a representation of her developing relationships with people. By locating herself (and her family) in the game of *Sim Town*, she articulated a kind of control over her position in the world. She could not control whether she was dropped off at the 5th D or not, but she could exercise control over her family avatars (iconic representations) in *Sim Town*. She could not control how her peers and the undergraduates treated her, but she could control who got placed in the game and how to represent them.

Sim Town affords the construction of worlds radically different from the technology-driven *Sim City 2000*. In *Sim Town*, neither financial planning nor the acquisition of money is an object. Players earn credits through doing good deeds. Instead of making budgets, players can create characters, give them names, choose their hobbies, create families:

> You can watch the Sims (the people who live in *Sim Town*) work, play and travel around town. You can peek into their homes and see where they live. You can even design your own special Sims—one in each of your towns—and decide what they like to wear, eat, play, say and what kind of pet they own. (*MAXIS Sim Town* Builders Guide, 1)

Instead of occupying the position of financial power by presiding as mayor, users are encouraged to pay attention to aesthetics of planning. Instead of allowing players an opportunity to learn about budgets, *Sim Town* encourages players to zoom in for closer views of the town, houses, and neighborhood aesthetics. In *Sim City*, clicking on any part of pop-up text windows (e.g., city newspapers) reveals text behind the headlines; in contrast, clicking on any section of the town newspaper in *Sim Town* simply closes the window. By excluding educational aspects of *Sim City 2000*, *Sim Town* de-emphasizes fiscal and civic inquiry skills and encourages players to pay attention to homemaking and the aesthetics of nest building such as laying lawn, shrubs, flowers, statues, and paths. Whereas *Sim City 2000* gives players a godlike gaze from outside the game space, *Sim Town* gives players a voyeuristic gaze from within. Players are encouraged to "peek into [Sim] homes and see where they live." With its focus on interior spaces, *Sim Town* resembles a feminized version of *Sim City 2000*. As Seiter notes, "Girls' play has always been depicted as closer to mundane—that is domestic—reality than boys' play" (1993, 76). In its emphasis on aesthetics, family relationships, and the decoration of characters, *Sim Town* resembles the world of doll houses that girls are thought to occupy. In this world, the characters are white, the families are traditional, and the economy is based on good deeds, education, and shopping.

In their expertise with different games, Freddi and Adam were exercising power—what I am calling masteries. Yes, each played with simulations in different frameworks—Freddi from the point of view of a character in the game asserting aesthetic paradigms, and Adam from the point of view of a "mastermind" outside the game trying to earn "credits." Nonetheless, beyond this notion that Freddi and Adam had insecurities, hopes, and desires that got played out in two CD-ROM games, their play was situated among networks of peers and adults struggling to define themselves within this "culture" of the 5th D. I am interested in these struggles as intersections between "fantasies," "realities," and the kids who made them. I focus on Adam's and Freddi's mastery over game worlds as a means to successfully position themselves.

Mastery over Others

In the world of the 5th D, Adam was renowned as a computer expert. Kids went to him for help, and Adam got to choose when, where, and to whom he would disseminate knowledge; this gave him access to a kind of power and control. Adam controlled knowledge in the economy of the 5th D. He would, for example, give his allies a secret code for unlimited funding in *Sim City 2000*. Given that locating the self in a powerful position in the economy can be seen as a kind of social coping strategy, Adam's and Freddi's developing mastery over game worlds can also be seen as an effective coping strategy—helping them master their positions in social space.

It was partly because of Freddi's inability to get along with her peers (and several adults) in the after-school world of the Boys and Girls Club that Freddi came to the 5th D at all. When pressed, Freddi would say that she came to the Boys and Girls Club because "she had to" and that she came to the 5th D because "there was nothing else to do." Despite her seeming lack of enthusiasm, Freddi was also a visible and expert player in the economy of the 5th D.

Freddi and Adam were similar in that they both struggled for (and over) power within the structure of the 5th D. They both got early exposure to the allure of being a YWA in a clash over power with each other. Freddi was playing a game on "Adam's *Sim City* computer" (the computer on which Adam kept his *Sim City* games loaded) when Adam challenged Freddi's right to play at a YWA computer. Freddi responded to Adam's challenge by pulling up one of Adam's *Sim* cities to show to her undergraduate partner. Freddi played until Adam came back over and started dictating how the game should be played. Freddi did what she wanted and ignored him at first, but Adam critiqued her by saying that she was building a house too far from her city. Adam told her that this was "stupid" because it was far from access to power and roads. An undergraduate described how Adam worked to nag Freddi and how she was unable to get power to the houses:

> He said, "Hey bimbo, do you see any power here?" Freddi rebutted, "Hey Adam, wake up and smell the coffee." They continued to argue even as Freddi finished her city and tried to save it. Adam had about 25 games saved on the computer and there was no more room to save another game so Freddi was angry. Adam went through his games to see which one he was willing to get rid of, but he liked them all and did not want to get rid of one, so he suggested that they ax another kid's game. Freddi started to build a new one, until the site coordinator came by and told them this game was not in the maze. (T.F., 2/9/95)

When Adam started telling Freddi and her partner what to do in the game and critiquing Freddi's city at their every move, this created a tension that gradually escalated into an intervention by the adult site coordinator. The site coordinator explained that *Sim City* was not in the maze and that Freddi must follow the maze.

Almost usurped by Adam's appeal to 5th D rules about YWA

territory, Freddi used this reprimand by adult authority as an opportunity to blame Adam for the disruption. The site coordinator told Adam to leave and Freddi ended up getting to resume play with her previous game. Thus, Freddi remained happily seated at "Adam's computer." Clashes of power such as these—between YWAs and non-YWAs, between boys and girls over games and computers—became opportunities for kids like Adam and Freddi to appeal to the rules as a means to assert themselves and power over peers.

A year later, when Freddi was also a YWA, she and Adam vied once again for power. In this excerpt, Adam stood behind Freddi, who was building the ritzy town of La Jolla in *Sim Town*. The following is their dialogue:

> *U:* Cool. [U = undergraduate (female)]
> *Adam:* How do you get credits?
> *Freddi:* For good deeds. What else, how else would you get credits?
> *Adam:* What's a deed? What's a deed?

Notice that Adam was orienting toward developing mastery of this new game in terms of his expertise with *Sim City 2000*. While he was unfamiliar with *Sim Town*, he saw similarities between the two games and was trying to impose the notion of "credits" that he was familiar with from *Sim City 2000* as a framework for understanding *Sim Town*. Freddi, seeming to know that she had the upper hand here, ignored Adam.

> *Freddi:* I need to get more residential areas anyway.
> *U:* Yeah.

Freddi's behavior suggested that she knew that she was in control of the game and that she possessed knowledge that Adam wanted. Adam, on the other hand, knowing that she knew something that he didn't, grew impatient.

> *Adam:* Well, how do you get a deed?
> *Freddi:* Do you need to know? You don't need to bug me so much.
> *Adam:* But how? How do you get deeds?
> *U:* Well, you build hospitals, and you build lakes, and you build fire stations.

Up to this point the undergraduate had been an accomplice to Freddi's display of power. She had not responded to Adam's queries. Nonetheless, she helped Adam with his question this time, which only increased Freddi's frustration.

Freddi: Why won't it [the cursor] move?
Adam: It's the dog. It stinks, you can't build it over the dog.

Adam's hypothesis infuriated Freddi, as if she perceived it as a challenge to the enactment of her powerful position. It was her game, her expertise, and "her" undergraduate. Why should she have to share these with another?

Freddi: I know, I've played this more than you have so, pssh! (to undergraduate): He's getting annoying.
U: Freddi!
Freddi: Make him leave me alone.

Freddi used her influence to get her way. Annoyed, Adam left, at least for a while.

Later, two other boys (S.B. and A.N.) tried to interject their own game logic. Again, Freddi stood up quite well for herself and defied the boys' logic of destruction and disasters.

S.B.: I like *Sim City* better.
Freddi: No, this is much funner, you don't have, you don't have a debt or anything.
S.B.: So, who cares. Debts are cool. Oh, disasters? What are the disasters?
Freddi: Nothing. There isn't a disaster.
S.B.: Yeah there are, look.
A.N.: Try disasters.
S.B.: Eco-villain.
Freddi: Nope.
S.B.: Can you look and see what the . . .
Freddi: They're all off.
S.B.: Just look at them. Go and look at them, see what they look like.
U: Maybe they'll destroy the whole thing?
A.N.: Yeah!
U: You want to see that, huh?
A.N.: No, you just look at it, like pull it down and you can decide which disasters.
Freddi: No, because there aren't any disasters. Look. (Freddi goes to disasters menu) All disasters off!

In this interchange, Freddi remained focused and in control. She did not waver in her resistance to these boys' logic, but instead remained committed to pursuing her own goals, confined to the aesthetic though they may seem.

Freddi created idealized game worlds to distinguish herself from others and their preferences. Bourdieu argues that people express themselves through their preferences: "Taste classifies, and it classifies the classifier. Social subjects, classified by their classifications, distinguish themselves by the distinctions they make, between the beautiful and the ugly, the distinguished and the vulgar, in which their position in the objective classifications is expressed and betrayed" (1984, 5–6). In the 5th D Freddi expressed herself through her relationship to others' tastes. When Natalie (N) (a tomboy who Freddi was always telling to get her hair done or wear a dress) interrupted Freddi, Freddi handled this disruption with mastery:

N: Now look, they're going to put a label. Ugliest dog in the world.
Freddi: Natalie, now really, how sensible was that one? I mean yeah, maybe it's funny sometimes but today? You don't want to know what kind of mood I'm in.

In this case, Freddi expressed mastery over her peer by formulating a quick reply to Natalie's jeer and by changing the subject back to herself and to her own interests.

(Freddi picks a girl character in the game and names it Freddi)
N: The dog—then name it Freddi.
Freddi: What? (Freddi chooses skateboarding as Freddi the character's main interest. Then she chooses a monologue. When I'm happy I say, "Can I go somewhere"; when I'm sad I say, "I hate you.")

Here Freddi used her character in the game to say what her character outside the game might have been chastised for. Freddi used her avatar to ventriloquate (Bakhtin 1981) what she did not dare say outside the game. She also used her character to express her hobby (she had just received a skateboard for her birthday) and what she would have preferred to do—go somewhere. Freddi located herself within the games, perhaps as a kind of escape. She used the games as a kind of protective shield behind which she could create her sense of identity.

Similarly, one undergraduate observer speculated that Adam was hiding behind the game. She remarked (in her field note) that Adam's hearing was very selective. She explained how she tried to engage

Adam by complimenting the efficient design of his city and telling him that only a very intelligent young person could accomplish such work. Adam then asked her if she really thought he was smart. When the undergraduate responded, "Yes," Adam explained to her that he did not think he was intelligent because he had been held back in school. In her field note reflections, A.E. explained how her role in this session had been to help Adam interact with other kids and that an emergent goal for her had been to have Adam teach someone else how to play *Sim City:*

> I noticed instead of interacting with other kids his age, Adam avoids them by playing *Sim City*. . . . This game allows him to realize that he too is smart. . . . Can the wizard write an [e-mail] letter to Adam encouraging him to play a game other than *Sim City*? He hides behind this game. It gives him a sense of security. Though we want Adam to feel secure, this security should not be based on a game. (A.E., 2/2/95)

While the undergraduate's observation that *Sim City* helped Adam to feel smart is salient, her notion of Adam avoiding relationships with kids is too simple. Adam did not hide. Rather, he used his *Sim City* expertise as an alternate means of relating to others. Rather than perform the role of outcast (subject to jeers like "When Adam swims, everybody leaves the pool because it is so disgusting" [K.H., 4/23/96]) or the role of kid with ADHD, Adam got to perform the role of *Sim* expert; in so doing, he gained mastery over his position in the 5th D world.

In the undergraduate descriptions of kids' play we see an underlying tension between what kids are doing in the 5th D and how the undergraduates frame what they see. Undergraduates play ambiguous and flexible roles in the structure of the 5th D. While they do not exist to discipline the kids (the wizard does that), they are there to help kids follow 5th D rules and guidelines. Since most of the rules are not explicitly outlined in the constitution, undergraduates make evaluations and assessments along the way. Carrying out their roles as participant-observers puts the undergraduates in positions of power, and where power is involved there is usually contest. Such struggles include questions of who will play what game and when, as well as who controls the mouse and who is "teaching" whom. For both Adam and Freddi, what counts as teaching was very much at stake. In a world where accomplishment was rewarded, even in the context of 5th D game play, these kids were understandably concerned about getting credit for their accomplishments. Nonetheless, they were sophisticated, and used their knowledge of the boundaries between acceptable and unacceptable behavior. This section deals with conflicts between adults and kids, particularly how Freddi and Adam used their expertise with 5th D games to gain power and control over supervisory adults.

Adam used his *Sim City* expertise to mediate his relationships with adults as well as kids. Adam developed strategies for dealing with supervisory adults and the "lay theories of development" that they used to determine what constituted appropriate behavior for him. By lay theories of development, I am referring to David Buckingham's research in *Moving Images* (1996) on parents' and children's reactions to television. David Buckingham found that parents often deployed "lay theories of development" to help define appropriate television habits. Lay theories tended to be predicated on factors such as children's age, gender, and maturity—children's ability to appreciate differences in the appropriate contexts for particular behaviors (68). It is precisely these concerns that Adam learned to mediate. Just as Buckingham found that children learned to employ a wide variety of strategies to regulate their own television experiences (254), Adam learned to regulate his own *Sim City 2000* habits. His expertise enabled him to define when, how often, and to what ends he played. When undergraduates objected, he knew how to position himself within the 5th D to take advantage of the rule structure.

Adam learned early on to use the goals and rules of the 5th D to earn more privileges. He learned to articulate his goal to become a YWA as a means to gain access to the YWA game *Sim City 2000;* as a result, he was allowed to keep playing *Sim City* even before he became a YWA. When confronted with the ways that he overstepped boundaries, Adam was expertly defensive. He appealed to the 5th D goal structure: "But I helped make the Task Card, we were making the task card yesterday" (E.C., 11/29/94). As a consequence of this appeal, Adam proudly bought himself enough time to have his city achieve "metropolis" status. In defense of his behavior, undergraduates explained how desperately Adam wanted to become a YWA. They reported pleas made by Adam such as "I can't wait until I am a YWA; I wish I were YWA" (J.P., 2/2/95), but then ended up describing how Adam's play with *Sim City 2000* prevented him from accomplishing that goal. Eventually adults caught on. It became lore in the 5th D culture that Adam was allowed to play *Sim City 2000* only *after* he had completed other tasks.

Three years after completing the requirement to become a young wizard's assistant, Adam was still being chastised. This time, adults suggested that he was not fulfilling his obligation to help others learn the games at which he was expert. In one case, the undergraduate observed another boy watching Adam play *Sim City 2000*. She noted Adam's methodical attention to detail but remarked that he didn't talk much unless Carl asked him a question (D.A., 10/12/95). Later that day, Adam quietly "taught" an undergraduate *Sim City* until he was interrogated by the site coordinator about his teaching methods. She told him that "teaching consists of verbal instructions accompanied with letting the other person play the game." Perhaps Adam felt ambivalent about the ways kids who teased him elsewhere used him as "a sort of tool or reference guide" (C.M., 2/13/95). Some field notes suggested that he used his knowledge of *Sim City 2000*'s secret funding code as a kind of bargaining chip for gaining friends and influence—strategies that he often had less success with outside the 5th D.

I am suggesting that when Adam entered the 5th D he was "virtually" escaping from the adult rules and authority that bound him outside the 5th D. Imagine that Adam's mastery with *Sim City 2000* emanated from a ten-year-old boy involved in a behavior modification program at home and school.[3] In Adam's world, his mother was explicit that she did not "want him sitting in front of the computer all the time" (D.G., 5/8/95). She also said that Adam had "become intimidated by sports because of the attention required to participate," but did not see the strange irony—that her kid diagnosed with ADHD would sit so focused and intent on his computer activity. When we contrast his mother's comments with the reception and admiration that he received at the 5th D, we see that while he immersed himself in game worlds, he was defying his mother.

Imagine a ten-year-old boy seated at a computer surrounded by people: an undergraduate student, several kids, and a graduate student videotaping his play with *Sim City 2000* (video shot on 10/2/95). The computer screen shows a technologically advanced city, complete with a population of over seven million people and over $27 million in available funds. Add to that the drama of disasters descending on his city and suggestions coming from all the kids and adults around him. In this picture Adam is in complete control and is not the overweight, ADHD "dork" described by kids and adults alike. Sections of his city are burning, and he alone gets to control which sections burn and which sections are saved. Different kinds of advice and questions are coming from those around him: "Build a nuclear plant." "No, build fission." With a determined tone in his voice, Adam responds, "Let's build the one that is most expensive."

With confidence and ease, Adam answers the questions he chooses to answer and ignores the others—even when they are directed by adults. A research professor approaches and reminds Adam to help the undergraduate—both by instructing her and by letting her have a chance to play the game. Adam is skillful. First he acknowledges the research professor and starts to stand up, but as the researcher turns away, Adam sits down again and tells the undergraduate that he just needs to "see what's left of his city." The undergraduate asks again if she can play. Adam, standing again, tells her to wait just a second because he just wants to blow things up. The undergraduate then turns to a friend and explains that Adam "wants to blow things up, as much as possible." Adam sits down again while two more boys gather around the game. Five minutes pass, and Adam is still at the mouse, narrating his spectacle before an audience of onlookers.

Freddi, unlike Adam, could be good about giving explicit instructions, but it was her attitude toward the other kids and adults that got in her way. Not well suited to containing her opinions, Freddi was likely to comment on people's intelligence, dress, weight, hairstyle, and so forth. She was described by undergraduates and kids alike as harsh and superficial:

> Freddi was very knowledgeable about the college system. She knew what midterms were. The fact that she had older sisters aided her knowledge. She was very advanced and mature for her age. Though she was very intelligent and wore nice clothes, she did not have any friends at 5th Dimension. She always played in the computer lab and never spoke to any of the other girls who were there and about her age. I think the thing that has constrained her from making friends is her harsh criticism for people. She is very selective about who is cool and

> does not talk to anyone she believes is a "dork." Though her basis for selecting people is solely on appearance and style. I wonder if she knows what personality is. (K.G., 4/2/96)

Freddi was not portrayed as nice, kind, or helpful. She was confident and assertive about her position in the world, and not just in terms of fashion. She knew what she had in her future—"to be a lawyer, politician, or President" (11/20/95). Is it this confident and assertive attitude that cemented others' distaste for her?

Like Adam, Freddi was comfortable with challenging 5th D goals. In the 5th D she was assertive about what she thought. In one example, an undergraduate explained how she e-mailed the wizard to complain about an all-boys day in the 5th D:

> When I first walked in before I wandered around the boys and girls club, I noticed Freddi sitting at a computer writing a letter to the wizard. I stopped to talk to her when a boy came up to her, I think his name is Steve. He told her that she couldn't be in here because she was a girl. He was telling her this in a "ha ha" sort of manner. Freddi got very defensive and told him that she was only e-mailing to the wizard and that the boys were allowed to do that on girls day so she should be allowed to on boys day. Steve then started to read her letter and Freddi got a big smile on her face. Her letter said "Dear Wizzzz, I think you made a big mistake in making an all boys day because I am one of the only girls to come here and everyday is an all boys day. (T.F., 3/14/95)

In that Freddi also wrote to the wizard to complain about game task cards, petition to play certain games, and complain about undergraduate practices, her behavior earned her a bad reputation. Undergraduates complained about her in their field notes and kids wrote letters about her to the wizard.

Ultimately the wizard intervened to mediate this conflict over power—the power for Freddi to act as she chose and the right of others to protect themselves from her insults. At stake were all the explicit and implicit rules of behavior. Could Freddi be convinced to act in a manner more befitting a young wizard's assistant? Procedurally, Freddi deserved to be a YWA, but her lack of respect for others defied an implicit but widely held assumption in the 5th D. The wizard responded to Freddi's disrespect by asking her to write an essay about respecting "diversity." With the help of two undergraduates, she wrote the following:

> We are all different in many ways. Such as our skin color, facial looks and expressions—also hobbies and interests. We have different friends and people to look up to. . . . I would probably be different than someone in Japan and look, think and act differently because we have different cultures and live in a different environment. Our environment and culture shapes our beliefs and who we are. At schools like Skyview, people wear different clothes and have interests in different things and people. There are different friends and friend groups, like one is cool boys and girls from different classes and in other classes it is different. Some people might shop at 540 and others at Wet Seal.

Freddi responded to her peers' concerns—much aided by the adults around her—without having to admit that she had been unfair to anyone. In this way Freddi passed the wizard's test and demonstrated her mastery of adult argument.

In the context of such struggles over power, Freddi's skill in handling relationships with supervisory adults emerged. She had to be able to manage the tension between adult expectations and her desire to play as she pleased. Consider the following interchange between Freddi and an undergraduate. She had just succeeded in getting Adam to leave the vicinity and was in the midst of designing a neighborhood in the *Sim Town* format:

> *Freddi:* The good thing about this one is you don't have to waste any money.

This was an explicit reference to *Sim City 2000,* where budgetary constraints can limit the kinds of cities produced. Since Adam was an expert at subverting these constraints, perhaps this was an invalidation of Adam's expertise as well Freddi then expressed relief that Adam had gone.

> (Freddi lays grass)
> *U:* Oh, this one's not for money? It's a grassy little area. I wish it would move along as you move it out, you know?
> *Freddi:* Thank God Adam's gone.
> *U:* What?
> *Freddi:* Thank God Adam's gone.

Again Freddi was successful in insulting Adam in the presence of an adult. While this was not a flagrant violation of any explicit 5th D rules, Freddi had nonetheless pushed the boundaries of what was acceptable and unacceptable behavior in this context. She succeeded because the undergraduate did not call her on this, but instead changed the subject.

With Adam out of the picture, Freddi asserted herself through her deployment of her aesthetics within the game:

(Freddi zooms out)

Freddi: How does it look? Now, how does it look?

U: It looks good, it looks good. Put some houses.

(Freddi puts down more houses) Oh that one's pretty. Yeah, that one. That, yeah.

Freddi: I'm going to make this Snob Hill, this one. Where all the (makes OK sign with fingers) houses are. I think the Bungalow would be pretty expensive. Have you ever been to the Bungalow?

Freddi expressed an attitude toward class position through her articulation of the space Snob Hill in the game. Freddi could have been parodying the upper class, but given the way she had consistently allied herself with the upper class (here, knowing that the Bungalow is an expensive restaurant), I would say that she seemed to be using the CD-ROM as a safe place to articulate aspects of her own identity.

Bourdieu's notion of the habitus helps me see the ways subjects appropriate representations and traditions. The habitus is a transformation of symbolic forms, not simply reproduced, but selected and reinterpreted from an available stock. This is how I see kids' play in the 5th D. I see Freddi reproducing tidbits of her class and social upbringing in forms as varied as her relationships with other participants in the 5th D; the genres of games she preferred; and the kinds of "worlds" that she produced. Her tastes were not uniform reproductions but were negotiated in relation to her conceptions of self and others. They were constituted in the social world through practices.

U: Oh, that one's cute, huh?

Freddi: It's a haunted house, you can't put it in.

N: What is that stupid castle thing?

Freddi: I don't know.

U: It's a mobile home. Uh, you don't want to put it in a ritzy area, do you? Cool hut.

Freddi: That would look kind of funny.

U: Uh, it looks kind of weird in there.

Freddi: So, there's some weird houses. Look at the houses we have. Look.

N: My house looks like that.

Notice how the others gave Freddi input according to their own preferences. She did not act on their suggestions, however, and continued to impose her own aesthetic detail, a logic informed by her notions of class and neighborhoods:

> *U:* OK, you want to put some more houses? All right. Why don't I put . . . oh stop, stop, oh.
> *Freddi:* Can I . . . here, let me try it.
> *U:* Can I please? Please.
> (Freddi takes mouse)
> *Freddi:* Please, please. I need more roads anyways.

Even after the undergraduate pleaded to participate, Freddi was confident in the logic of her position and loathe to give away any control.

Accepting some defeat, the undergraduate resigned herself to giving Freddi advice. The undergraduate appealed to the logic of economy within the game and suggested that Freddi build an apartment building to handle the housing shortage.

> *U:* Gee, an apartment building would fit a lot of people. Why don't you build three or four apartment buildings on a block? They'll fit a lot of people in.
> *Freddi:* There's one right there.
> *U:* Yeah, see that's an apartment building right here.
> *Freddi:* Yeah, but who's going to want to live right next to an apartment building? You can't put it in. It's too big.

Acquiescing a bit, the undergraduate championed her suggestion once more.

> *U:* Well, put it like here along the road, you know what, there's people with less income. There's people that work in the video arcade. You know, they don't have money to afford a house.
> *Freddi:* Oh, that's cool, yeah. I mean who of the people that can afford a house are going to want to see one of those around? (pointing to the apartment building) Oh, they're always by schools.

Unconvinced by the undergraduate's ideas, Freddi reasserted her class-based logics. Notice how confident Freddi was about her orientation to class. Was it how Freddi asserted her aesthetic preferences that made Freddi so dislikable—or that she was from an upper-middle-class neighborhood?

Bourdieu's notion of the habitus is both symbolic and structural.

Environmental features produce habitus, and habitus is produced in the structuring of practices of representations:

> In fact, through the economic and social conditions which they presuppose, the different ways of relating to realities and fictions, of believing in the realities they simulate, with more or less distance and detachment, are very closely linked to the different possible positions in social space, and consequently, bound up with the systems of disposition (habitus) characteristic of the different classes and class fractions. Taste classifies, and it classifies the classifier. Social subjects, classified by their classifications, distinguish themselves by the distinctions they make, between the beautiful and the ugly, the distinguished and the vulgar, in which their position in the objective classifications is expressed and betrayed. (1984, 5–6)

The way Bourdieu theorizes that social spaces transmit class as positions to occupy seems particularly salient for understanding Freddi and the positions she occupied in the 5th D. It also helps me understand why I both disliked and admired Freddi.

Despite my antagonism toward Freddi's elitist preferences, I found myself admiring her expertise. Unlike those who engage in so-called typical girl behavior, Freddi developed useful debating skills more often associated with boys and men. Carol Gilligan (1982) describes the ways girls are socialized to think in terms of responsibility and care: "Thus, in all of the women's descriptions, identity is defined in a context of relationship and judged by a standard of responsibility and care. Similarly morality is seen by these women as arising from the experience of connection and conceived as a problem of inclusion rather than one of balancing claims" (1982, 160). It is hard to imagine a description less applicable to Freddi than this. Since Freddi developed argumentative skills (as well as technological expertise) that were deployed both in and outside the 5th D, her behavior defies all accounts of girls' "naturally" gendered positions in the world. I learned to appreciate Freddi's strength and see similarities between her precocious behavior and my own childhood relationships with people. As the new kid at school (seven times), I remember defending everything from my choice in clothes to my parents' lifestyle to the kind of books I liked to read. I know that I distinguished myself from the "immature" kids around me and spent lunch hours in the classroom. I also remember having been described by adults as opinionated and argumentative. I find myself wishing for more contexts in which young girls are encouraged to stand up and argue with the adults around them.

I find myself thinking back to the time when I was sent to detention for arguing with my second-grade teacher about the interpretation of Friday's reading assignment. He had told us to read the first chapter of our new reader and answer the questions at the end of the chapter. That weekend I immersed myself in the world of the reader. Not only did I read all ten chapters, but I answered approximately a hundred of its questions and proudly handed them in to my teacher on Monday. I can still recall my devastation when the teacher sent me to detention for (1) disregarding his assignment; and (2) debating with him about why he should accept my work. I can identify with a skillfully focused albeit headstrong Adam—struggling to do something he enjoyed in a world structured by others' rules.

Giving Adam and Freddi opportunities to challenge themselves and the adults around them created new opportunities for these kids and new positions worth occupying. Consider Walkerdine's supposition that

> What seems to be at issue is not a series of roles or simple identities or images which are fitted to girls. Nor is it a matter of certain roles being "stereotypically feminine" and therefore allowed, and others not. Rather, we need to understand the relationship between those practices which not only define correct femininity and masculinity but produce them by creating positions to occupy. (1990, 103)

We need to understand the conditions under which nontraditional practices can create new opportunities for girls and boys. We must ask why is it that when Adam displayed a keen and sustained focus on *Sim City 2000*, his goal-directedness got constructed as a problem in the context of the 5th D, and how is it that when challenging peers, practices, and adults alike, Freddi also got constructed as a problem. We must wonder how despite being constructed as problems, both surpassed their peers by demonstrating expertise with complex simulation games. The appropriation of empowering roles is complicated; this makes Adam's and Freddi's movement between mastery of fantasy game worlds and real-life situations more interesting.

My reading of Adam's and Freddi's play suggests that we query the relationship between fantasy and mastery. We know that outside the 5th D Adam was teased by peers and criticized by adults. But in

Sim City 2000 he was an expert; he used his expertise to strategically reposition himself in the social world of the 5th D. Likewise, Freddi's critical remarks got her into trouble with her peers. In the context of computer gaming in the 5th D, however, she articulated these criticisms as aesthetic difference and built them into substantive arguments. According to Walkerdine, "it is specifically overt challenges to the teachers' authority [and] power over knowledge, which are validated and hailed by teachers as evidence of real understanding or 'brains', 'brilliance', and so forth" (1988, 209). Thus Adam's and Freddi's challenges to authority can be interpreted as a kind of mastery over real-life situations. They used the expertise they developed in their mastery of fantasy game worlds to reposition themselves in everyday contexts beyond game worlds.

Future Worlds

Adam's and Freddi's play with CD-ROMs in the world of the 5th D provides for an opportunity to study *Sim City 2000* and *Sim Town* as tools to mediate and remediate one's position in the world. The 5th D is the kind of place where kids go to accomplish different sorts of things that they may not always be able to accomplish elsewhere. Thus, the nuanced negotiation of kids' identities via their play with these CD-ROMs has different meanings in the 5th D than it would in other social spaces such as the home, classroom, or playground. In this context I have described multiple and often interwoven types of mastery. Now I propose a three-level schema for conceptualizing these masteries: (1) self-concept (mastery of game worlds); (2) social world (power over others); (3) future worlds (effects that will have consequences in other spaces).

First, there is the mastery of game worlds and game contents. I do not mean to conflate knowing how to do something with knowing the contents of that something. Learning how to do something well, like playing a musical instrument, requires practice; this is different from knowing the contents of something, like musical notes in a score. In most activities, like music, mastery of the practice and mastery of content domains are closely intertwined. This is also the case with mastery over game worlds. In order to build a successful city/town, one needs to know something about its elements and how they fit together. Usually this takes practice (perhaps a bit less practice, the more one already knows). I argue that since kids gain confidence and skills by developing mastery of CD-ROM game worlds, their mastery of their fantasies in and around CD-ROMs gives kids empowering tools that they can employ elsewhere.

Given that Adam and Freddi were not born knowing (or wanting to know) how to play *Sim City 2000/Sim Town*, we can address the relationship between game worlds and outside worlds. While I can not claim access to Adam's and Freddi's subjective worlds, I do have access to their representations of themselves—in talk, writing, and games. Adam and Freddi expressed pleasure in their mastery over CD-ROMs in the context of a social world, the 5th D as haven. They used their involvement with CD-ROMs as a means of dealing with others in the social world, and this is represented in their dialogue with peers, adults, games, and the structure of the 5th D. Even as an escape from peers, their play revolved around imagined others and helped them develop expertise that earned them accolades and freedoms beyond those given to their peers. This freedom gave them license to exercise power over others—even to subvert the dictates of supervisory adults. Adam and Freddi developed a feel for agency and how to mobilize their wishes in the social world.

With agency they gained more knowledge and a taste for the pleasures of mastery—over people and things in the social world as well as over their position in that world. Adam and Freddi learned about the effects of power. They discovered that exerting power produced consequences for others and that under certain circumstances they could position themselves to be better able to exert that power. Adam and Freddi learned to deploy knowledge of technology, rules, and procedures, and in learning how to manipulate those tools, they developed skills that would enable them to display power over peers and adults. Adam and Freddi learned how to position themselves in relation to a given problem space so as to be better able to transform a given social reality into one more befitting their tastes and desires.

NOTES

1. In conjunction with the Laboratory of Comparative Human Cognition (at UCSD) and the Institute for Research on Learning (in Palo Alto), I attended

the 5th D and its sister project La Clase Magica as a fieldworker during the 1995–96 school year. Data collection duties were distributed among several people; we videotaped nearly every day, interviewed participants, took notes, made maps, and so forth. Katherine Brown and Mizuko Ito deserve special mention for their hard work, expertise, and support. Please send comments to vgack@weber.ucsd.edu.

2. Many subsequent field notes describe how Adam brings "junk" rummaged from trash cans into the 5th D and uses this space to engineer new things—toy airplanes, motors, lightbulbs, and so on. Adam would then take these creations home and to school.

3. In this system, his mother and teachers give him tasks to complete, and for each task that he completes, he earns credit toward participating in an activity that he is interested in—such as playing on the computer.

WORKS CITED

Bakhtin, Mikhail M. 1981. *The Dialogic Imagination*. Ed. Michael Holquist. Austin: University of Texas Press.

Bourdieu, Pierre. 1984. *Distinction*. Cambridge: Harvard University Press.

Buckingham, David. 1996. *Moving Images*. Manchester, UK: Manchester University Press.

Davis, Madeleine, and David Wallbridge. 1981. *Boundary and Space: An Introduction to the Work of D. W. Winnicott*. New York: Brunner/Mazel.

Deri, Susan. 1978. Transitional Phenomena: Vicissitudes of Symbolization and Creativity. In *Between Reality and Fantasy*, ed. S. A. Grolnick, L. Barkin, and W. Munsterburger. New York: Jason Aronsen.

Gilligan, Carol. 1982. *In a Different Voice: Psychological Theory and Women's Development*. Cambridge: Harvard University Press.

Ito, Mizuko. 1997. Interactive Media for Play: Kids, Computer Games and the Productions of Everday Life. Ph.D. diss., Stanford University.

Jenkins, Henry. 1992. *Textual Poachers: Television Fans and Participatory Culture*. New York: Routledge.

Seiter, Ellen. 1993. *Sold Separately*. New Brunswick: Rutgers University Press.

Tulloch, John, and Henry Jenkins. 1995. *Science Fiction Audiences: Watching Doctor Who and Star Trek*. New York: Routledge.

Walkerdine, Valerie. 1988. *The Mastery of Reason*. London: Verso.

———. 1990. *Schoolgirl Fictions*. London: Routledge.

Chapter Ten

Showing and Telling

Developing CD-ROMs for the Classroom and Research

Leslie Jarmon

Grounded in my own experience using CD-ROMs in research and in the classroom, this essay examines two academic practices that have been changed by CD-ROM technology. First, presenting instructional material in our classrooms and at scholarly meetings is an important part of academic activities, and using CD-ROM technology has enriched the collaborative nature of these kinds of interactions in specific ways. Since the technology allows me to show and tell in new ways, the interaction between me, the students, and the content material has changed. Second, many educators spend a great deal of time handling data in various forms; we collect, store, organize, manipulate, represent, and disseminate data as we conduct our research and teach our classes. CD-ROM technology has enhanced my handling of data in several concrete ways that I believe are relevant to other teachers and researchers across many disciplines. Some of the most powerful enhancements include being able to create and manipulate new dynamic representations, to locate text and multimedia alongside one another, and to have very rapid access to these digital forms of information. This essay discusses how using CD-ROM technology has changed these two academic practices for my teaching and research; and it also identifies the specific properties of CD-ROMs that have provided for these changes.

The Temporal Dimension

To introduce the discussion of my classroom and research experiences, I want to offer a brief account of how I came to use CD-ROM technology. My training as a scholar and a teacher, like that of many academics, has relied heavily on the use of printed texts, lecture notes, journals, printed syllabi, handouts, and, of course, student essays, tests, and other written materials. My thinking about how to do research, how to prepare for a class, and even what to do once in class has largely centered around written texts, class discussion about them, and written assignments. This changed as I began to learn more about digital technology and about CD-ROMs in particular.

Many of us study and teach about phenomena for which time (and events unfolding over time) is of critical importance to our understanding. For example, my own field is communication; I study and teach about what kinds of interactional tasks are getting done when, for example, two people carry on an everyday conversation over a relatively brief time span. So a problem in my field has always been how to represent in text the dynamic, embodied nature of a conversation. But other disciplines share the problem of representing temporal aspects of a dynamic process using text: consider processes like chemicals reacting with one another, tumor cells dividing, stars exploding, stress increasing on an iron beam, continental plates shifting, a hybrid rice plant growing, a tropical storm self-organizing, a dance troupe performing—all these various phenomena share at least two important features related to time.

First, they are dynamic processes whose very identification in our world entails a reconfiguring of their forms over time; they unfold and change through time in ways that are structurally recognizable (the chemicals combine into a new compound with a new form; the tropical storm develops into a scale-three hurricane, its shape transformed; the increasing stress generates tiny fissures in the weakening steel beam; the growing tumor encloses and assimilates nearby tissue and changes shape; the friendly conversation grows into a recognizable argument; and so forth). Representing in text these changes of form and structure as they occur through time has often been a cumbersome challenge; one needs an inordinate amount of text simply to "tell" the story of change. Furthermore, for some of these subjects, representation of any kind also requires a manipulation of our conventional treatment of time itself: to describe the process that is unfolding, we may have to speed time up or slow it down. So there has been something unsatisfactory about our textual representations of dynamic phenomena.

A second time-related feature of a variety of subjects we teach about is that there is a sequential structure to the very nature of the phenomena; something happens and then something happens next and then something happens next, in a particular (and often critical) order. Researchers, teachers, and students often attempt to focus their exploratory attention on the details and underlying structures of such emerging processes across time.

What do dynamism and sequentiality have to do with using CD-ROMs in the classroom and in research? To oversimplify, using this technology I have been able to make available to students (and to other scholars) representations of the dynamic structural and sequential aspects of a particular communication pattern by being able to situate multimedia (visual, aural, moving) examples of the pattern alongside the analysis in the written text. That is, with the CD-ROM my students can see dynamic representations of examples of the phenomenon in real time (or in slow motion, sped up, backwards, and so forth; more on this later). This has had a tremendous impact on our interactions, and there will continue to be repercussions throughout the research community as we are increasingly able to offer multiple representations to our students by means of this technology.

In this essay I describe how I have been able to use CD-ROM technology to bring text and moving-image representations (digital video or simulations) together in one place for my students. Also (and on another level) I want to describe how using CD-ROM technology has extended the boundaries of my classroom and has simultaneously enriched student interaction with class materials. Although my own experience has led me to be an advocate of CD-ROM technology, there are some difficulties with the technology as well, and some of these issues will be addressed. But the focus of this essay will be on the ways its usefulness emerged from and for my own academic practices. Let us begin with the classroom experience and other academic presentational activities.

In the Classroom (and Out): New Kinds of Interaction

On the first day in class with the CD-ROM, the seminar students crowded into my office and we sat bunched together in front of my computer screen. The CD-ROM we were going to use is one I created for my Ph.D. dissertation (Jarmon 1996), a micro-analysis of how it is that humans communicating in everyday situations are able to make sense of what we do: not just with our talk and gestures, but with our faces, our whole bodies, and our physical orientations to one another. The CD-ROM demonstrates and analyzes how humans rely on minute structures of movement, timing, and repetition, among other resources, in producing communicative actions in face-to-face interaction.

I had just handed out their copies of the CD-ROM. Two students gingerly held the case as if it were an art object; another studied the cover, opened the "jewel-box," and lifted out the CD with a knowing hand. The students ranged from those who were barely computer-literate to "hot shots"; but no one had used an interactive CD-ROM before. Their homework assignment would be to "read" several sections of the CD, and I had planned to take about fifteen minutes to introduce them to the technology so they would know how to "open" it on a campus computer or at home. As we sat there crowded together, I assigned one student to be the "mouse clicker" throughout the demonstration (this provided empirical evidence that a peer could "do" it). Another student carefully loaded the CD in its tray, and I had her repeat the loading process for everyone to watch while I mentioned that it was a good idea to keep one's fingers off the playing surface.

This was their first experience with a multimedia CD, and I had forgotten what it was like to play with one for the first time. I realized right away that I would have to slow the demonstration down. The students' eyes were locked on the computer screen. A videotaped clip we had studied in class all week (using the VCR) suddenly appeared on the screen. Alongside it were a transcript, which they recognized from handouts in class, and some textual analysis. Then, as one student clicked the "play" button, the clip played itself "magically" in slow motion with sound, background music, and a voice-over. Several students started laughing, and someone shouted, "Play it again! Make it do it again!" What was supposed to be a brief introduction on how to get started ended up taking most of our class period. An hour later they were still engrossed in the CD and most had mastered the basics of how to "play" or navigate through it. They were talking to each other, ideas flurried around, and they were exploding with questions. I had anticipated some interest and some trepidation, but not this level of engagement. Would it continue? What difference might the CD make in their engagement of course content?

Over the next few days I had several calls and e-mails from students asking for specific help in navigation through the CD, or posing questions about the homework section, or comparing some aspect of analysis on the CD to their just-started individual research projects. This, too, was unusual in that the sheer frequency of student-initiated interaction with me increased markedly. Three examples of those interactions follow.

In this course students had to learn to use several kinds of technology, including video cameras, VCRs, and electronic mail. Early in the semester one woman articulated deep reservations about her technical capabilities, even in operating a VCR. Several days after our first in-class demonstration of the CD-ROM, she called me and asked if I could help her "get started" again. We arranged for her to call me while she was seated at her computer at home, and we walked through the process, beginning again with how to insert it in the computer bay. In short order, her insecurity with her own ability to interact with the computer changed into a new kind of confidence. A few days later, she called again to ask a very different level of question: the re-playable video clips on the CD really brought home to her key points in the analysis, and could she plan on doing something like that with her own research materials? She had made a leap from relating to the CD-ROM as something unknown and slightly terrifying to regarding at least a part of it as a model for how she might begin thinking of her own research project. The technology had not become completely transparent, but the ways to make the CD "work" in her computer had become transparent. It was now "becoming" a new kind of tool she was using to learn from in order to accomplish something else (her own project).

More surprises were to follow, but already I realized that these students had a keen interest in the CD-ROM technology itself (as well as the CD-ROM's content). The CD had already altered our interactions and resulted in an increased frequency of student-initiated com-

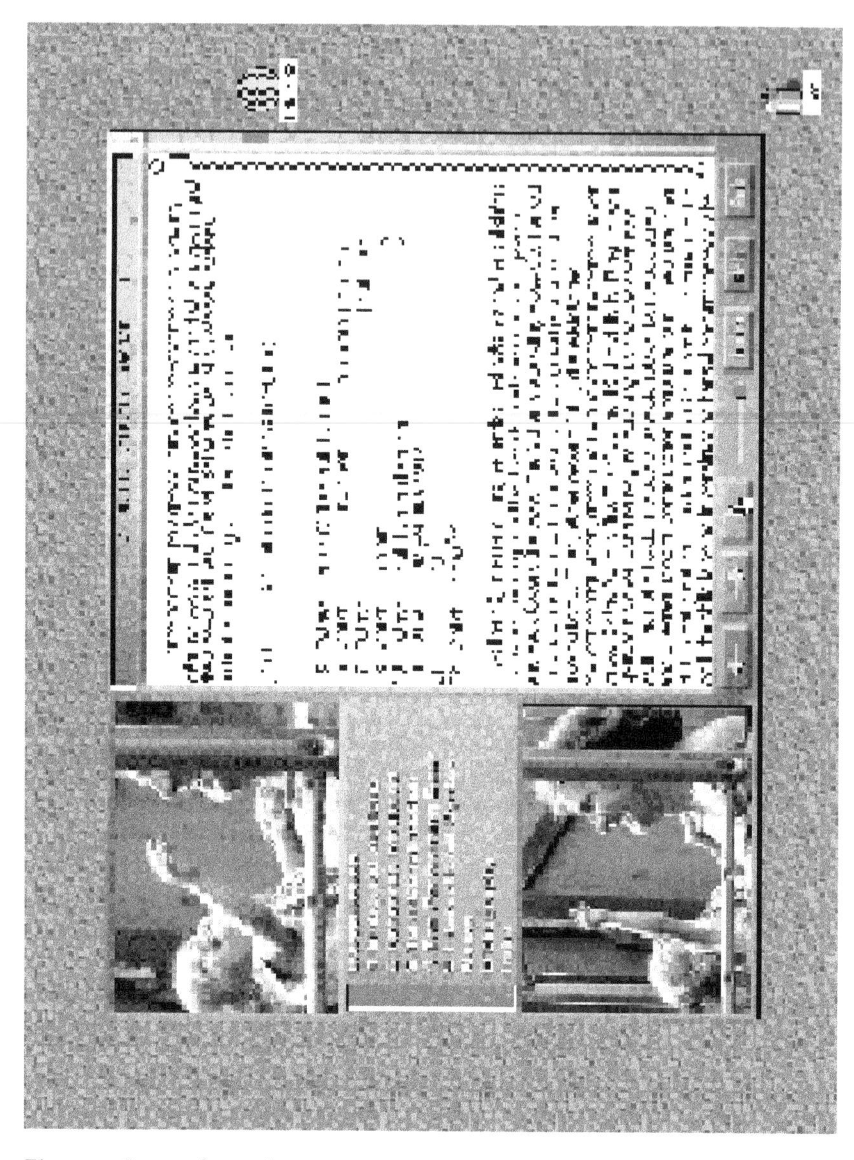

Fig. 10.1. Image from the computer screen running the CD-ROM for a communication course (Jarmon 1996).

munication, their increased attention to details of the content, and their heightened interest in the form and structure of the CD as a model for their own class reports.

But the use of the CD-ROM brought about an even more significant change in regard to our concept of a class: in a critical sense, the spatial and temporal boundaries of the class were extended beyond our room and our scheduled meeting time. In most classrooms (and, similarly, in conference presentations) one can use videotaped examples, lead a discussion, lecture about patterns visible in the examples, engage students' questions, and so forth. But classes and presentations all have time constraints. When the bell rings and class is over, the VCR will get used for another class, the precious videotape is stored away, and of course the lecturer is gone. Even when we arrange to make course videotapes available for students to view outside class, whatever guidance and analysis we as teachers voiced in class while showing a videotape will be missing (and, of course, copies degenerate the clarity of the images, and each playing damages the magnetic tape).

But with the CD-ROM my students get the video examples, the voice-over lecture, and the textual description and analysis—all right there together—time after time. They can replay the data "movies" and pause them on individual frames if they like, revise their notes, reread the text, reexamine the claims being made, connect the details of the analysis to the details of the examples, and challenge the findings—in ways that completely enrich the out-of-class interaction with course materials. Whether at home on their own computers or in the computer study labs located across the campus, the students themselves suddenly have access to some of the course's multimedia source data and analyses. The CD-ROM extends the temporal and spatial boundaries of "class" in these highly productive ways, adding to the "time spent" with course concepts and illustrations.

But this extended classroom also adds to students' engagement of more metaconceptual issues (for example, the structure and organization of claims and evidence). The extension of their learning into an examination of how a piece of scholarship is being presented has implications for their other classes as well as mine. It enhances their ability to attend critically to organizational and presentational issues across the curriculum.

There are difficulties with using a CD-ROM for a class. A basic prerequisite is for the students to have access to computers equipped with CD-ROM players, and this is clearly a challenge for many institutions. Furthermore, when and where a student can use the institution's equipment is frequently limited; and the CD-ROM is not something you can curl up with in bed. These are not trivial issues. If they are not addressed, CD-ROMs simply will not be very useful in the classroom in some of the ways printed texts can be.

Still, for those fortunate enough to have access to the necessary technology (and this number is increasing), our conventional notion of "class time" can be extended across conventional space and time constraints, and what gets appreciated as "content" can be extended as well into a consideration of how scholarship can be presented. In my class the students began to imagine ways they could effectively integrate various media into their own project reports. All of them would be making oral presentations to the rest of the class, and some were to appear on a panel at a national conference. These imaginings were not simplistic; rather, they reflected the students' new concern with the organization of their research presentations. Imagining how to organize their materials led them to ask critical questions in class about the logical arrangement of so much information—not only text, but still images and video samples as well. They raised questions about being able to insert freeze-frame images from their video data into the text of their papers. They also began to ask questions about creating their own digitally manipulated videotaped presentation that would highlight in a short amount of time the essential aspects of their analytical claims.

I, like most educators, am concerned that students learn to make cogent arguments in their papers by creating logical narratives supported by evidence and illustrated with specific examples. But the CD seemed to trigger in some of the students a much deeper concern with issues of organization and representation. They asked questions about dealing creatively with different forms of information for their presentations (text, images, and video), offered suggestions to one another, and sought more consultation with me in person and by e-mail. I believe that this kind of active engagement and learning about the persuasive organization of scholarship would be valued in any classroom. Our use of the CD-ROM helped students learn to think critically.

The final example of a new kind of classroom interaction is also related to how students were engaging the CD-ROM and involves another unexpected discovery on my part. It turned out that the usefulness of the CD-ROM for many of my students extended beyond the particular ways I had planned in designing the course syllabus. Teachers share a concern with methodological content as well as knowledge-content; we want our students to learn how a field of knowledge gets constructed and is practiced as well as what a field of knowledge can teach us about the world. As the semester progressed, most of the students' analysis skills were improving and their grasp of the larger research domain was increasing through course readings, class discussions, and their own studies. Suddenly, however, much later in the semester, some students began making frequent references in class to the CD-ROM as a model of how to present analytical claims alongside visual examples from their own source data. Again, in the CD-ROM, the digitized video examples, the voice-over lecture, and the textual description and analysis are all right there together, in one place. The students began to take conscious note of that model and emulate the richness of its representations.

In addition to asking questions about representation and organization, the students were engaging the CD-ROM in another way. As students began to discover recurrent patterns in their own research projects, the CD-ROM also served as a model for how to indicate increasing degrees of detail in those patterns. For example, on the CD there are numerous "analysis movies" of people talking; these play for a few seconds, after which one can play them again in slow motion to emphasize a behavior, and then again with a "zoomed in" view of the participants' faces or hands to focus attention on the details of movement. From such examples the students were continuing to learn how to do more detailed observation and analysis. This revisiting of the CD-ROM as a model for the practice of doing analysis—a "how-to" guide for the kinds of analytical tasks they now felt a need to do themselves—was also proving to be a reengagement of the content of the CD study.

I want to emphasize this final point about content because a key feature of the best interactive academic CD-ROMs is that they can offer tremendous depth of content information located in one "package." The primary audience for the CD-ROM I used in my courses was not students per se but rather advanced scholars in the field of communication. Accordingly, I had anticipated that much of the analytical material on the CD would be inaccessible to these novice scholars. But over the course of the semester, as many of the students acquired better-trained eyes and ears and increased their understanding of other research in the field, an unexpected situation arose: they spent more time engaging the analyses presented on the CD than I had anticipated. As they learned more about how to conduct microanalysis, the CD provided them with an interactive way to test their understanding of analytical claims: they could play and replay the video instances and simulations and reread the analysis alongside the examples to see how each analysis was done. I believe this ability to interact repeatedly with the video materials and analysis on the CD-ROM greatly enhanced my students' ability to do more learning on their own.

Furthermore, with the ability to view more closely and repeatedly dynamic representations, video clips, and simulations of the process under examination, the students were able to develop an experiential grasp of conversation patterns. Michael Polanyi (1967) referred to the experience of closely attending to something as indwelling; with the CD-ROM, students could see the speakers' faces and hands and hear their speech, closely and repeatedly. In other fields, for example, students could experience a speeded-up, close-up simulation of a rice plant unfurling its first leaves and could get a feel for the holistic process of plant development from a perspective that is very different from our normal human gaze. Or they could replay a slow-motion video segment of a bird in flight to experience the dynamic and rhythmic structure of wings working through air. Such experiences of indwelling in the phenomenon being studied can be prompted by the moving representations one can present on CD-ROM, whether videotaped or simulated.

By late in the semester, both in our class discussions and in our one-on-one sessions, the students were offering better analysis and better comparisons to patterns and details from their own data collections. Furthermore, since everyone in the class had viewed and reviewed the materials on the CD-ROM, we also now shared a common pool of specific data examples to reference, to argue with and against in class.

In sum, although I thought I knew the purpose for including the CD as part of the course (largely for its knowledge-content), I had not anticipated that it would end up serving so many purposes as a

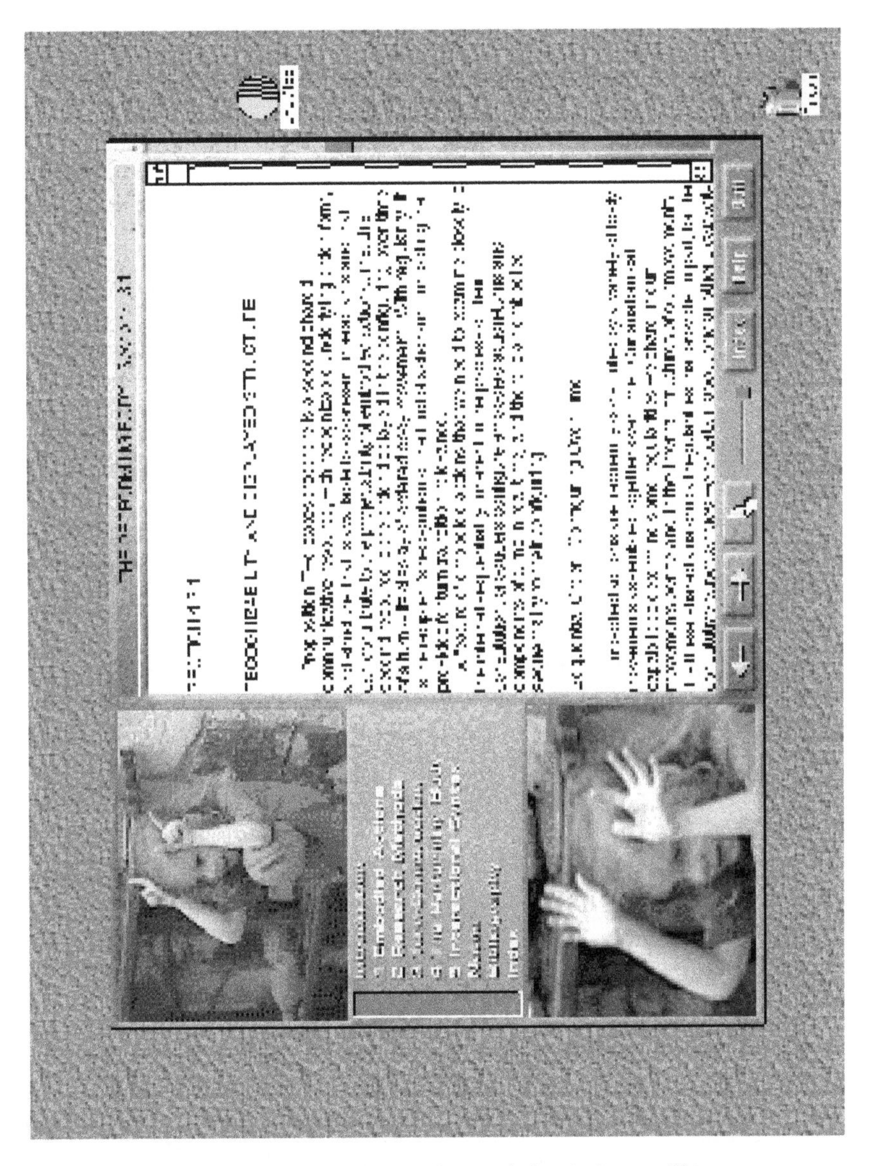

Fig. 10.2. The analysis movies play in the top left window, still images appear in the lower left window, and scrollable text fills the largest window.

teaching tool. I believe that CD-ROM technology can have significant pedagogical impact on the ways students interact with course material. For educators concerned with creating in-depth course content and generating opportunities for students to engage course material critically, this technology can be very powerful.

Showing and Telling: Together Again

In this section I will to look at some of the ways digital technologies, and CD-ROM technology in particular, have affected the way I deal with data. What has been crucial for both my research and teaching is the CD-ROM's ability to bring different representations of a phenomenon together "on one page." Bruno Latour (1990, 60) has observed that "if you want to understand what draws things together, then look at what *draws* things together" (emphasis in original). In other words, on a particular screen or "page" of a CD-ROM, my students were allowed to engage with more than a written analysis representing the phenomena of the conversation, more than a transcription of the conversation, more than a single frozen frame from a video of the conversation, and more than a digital video clip of the conversation (which includes, of course, the sound track of the people talking as well as the moving video images of their bodies, faces, etc.). Rather, students were able to examine and engage this whole collection of carefully crafted and very diverse representations of the conversation all together. This array of representations conveys a depth of information and detail, a richness of different perspectives on the phenomenon, a juxtaposition of representations of the conversation that are simply unavailable through any single form alone.

As suggested earlier, the study of any process that involves dynamic components can be enhanced when multiple representations are brought together in one space. In the case of my communication CD-ROM, the representations are aural, visual, sequential, textual (inscribed), and dynamic in nature; but the same would apply for other phenomena as well. Additionally, the digital video and computer-generated simulations used on this CD-ROM provide for what might be characterized as a six-dimensional representation:

Dimensions 1, 2, and 3: The images from the digitized analog videotape provide the conventional experience of three-dimensional representations. Similarly, for those researchers in other fields where there is no analogue videotape to begin with, computer-generated animations can create the impression of three dimensions to highlight dynamic aspects of the focus of study. Animated simulations are extremely powerful ways of depicting motion and shifting perspectives.

Dimension 4: The digital images, whether digital video or graphic animations, have duration and represent moving objects, adding temporality, a fourth dimension.

Dimension 5: The moving images can be played nonlinearly (stopped, restarted, in reverse); the conventional unidirectionality of time can be manipulated. This is a fifth dimension. Clips can be shown in slow motion or speeded up. Furthermore, conventional uniformities of spatial representation and the same sequence of images can be manipulated as well: one can view the "action" as though from a different angle, so that, in effect, the location of the viewers is shifted long after the fact of the original videotaping.

Dimension 6: The technology provides for a sound track as well, and this aural domain adds a sixth dimension to the enriched representations of dynamic phenomena. For example, a voice-over accompanying a clip of a cell dividing can provide key analytical information while the dynamic details of cell division are simultaneously unfolding in a moving image for the viewer. This is showing and telling, together, and it is available, immediately, for repeated viewing and listening.

There is an experiential difference for the researcher and student when engaging such representations of data and when conducting analysis in this information-rich "environment." With the addition of digitized video movies on the computer screen, the CD-ROM interface offers a view of something moving through time in space, changing. This is an interactive view that also affords access to details of the unfolding action; one can pause on a single frame, or rewind and see again something that came "before." A CD-ROM interface can provide tools to examine the phenomenon or the process being studied in different ways. A text window next to the movie window can simultaneously display written analyses, transcriptions, charts, still drawings, and more. Latour has observed that "A new interest in 'Truth' does not come from a new vision, but from the same old vision applying itself to new visible objects that mobilize space and time differently" (1990, 3).

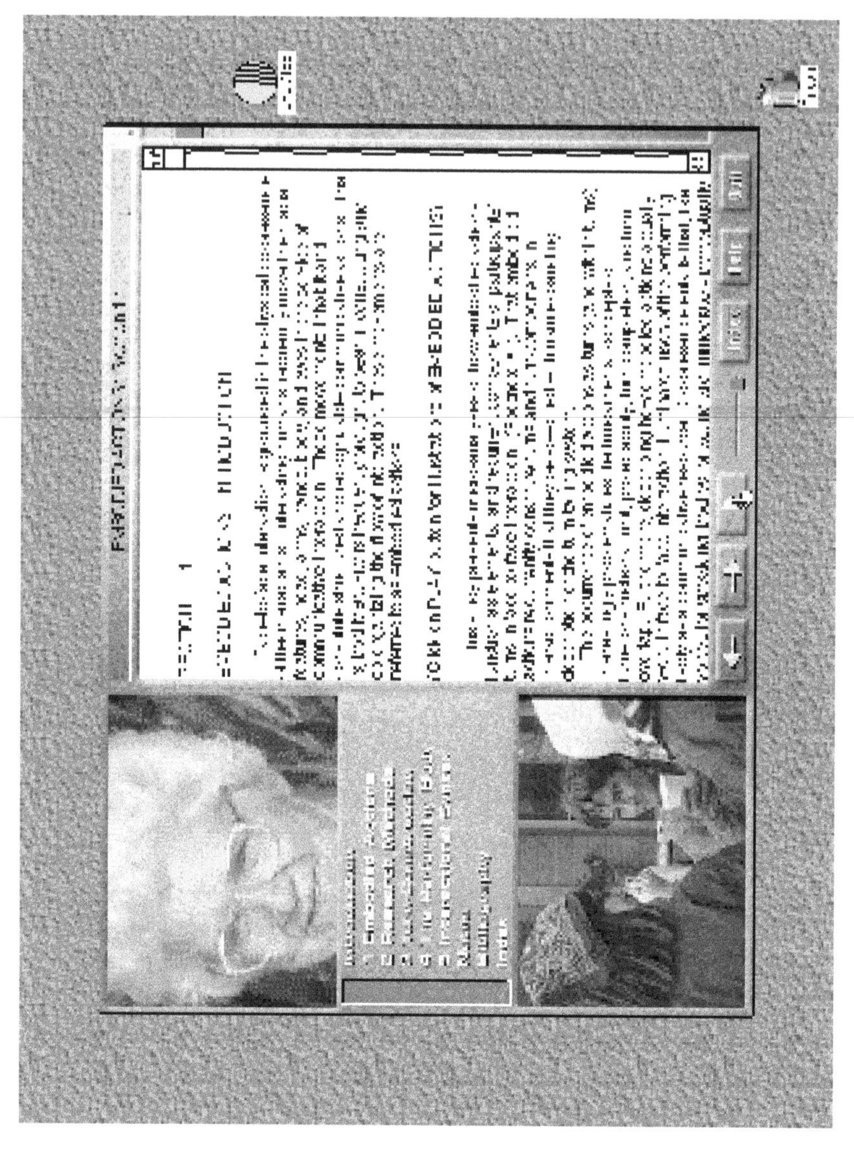

Fig. 10.3. The analysis movie that plays if you could click the play button shows a slow-motion segment of a woman talking; the image is zoomed in on her face for a "close-up" angle after the fact of the original taping.

CD-ROM technology allows us to take our data and create new visible objects and representations that mobilize "time and space differently" than did our simple text standing alone. Because an important part of what students must learn to examine closely is the moment-to-moment, bit-by-bit unfolding of a process in time and in space, our ability to mobilize and manipulate these two ways of representing experience (time, space) gives educators a new way to handle the phenomena about which we teach.

Thus CD-ROM technology provides for our rapid access to sight, sound, and movement. The CD-ROM brings nonlinear representations of visual processes into the student's own computer. In such arrays of representations, "you present absent things" (Latour 1986, 8); you bring the dividing cell, the growing rice plant, or the conversationists to the "text" and to the student. With multiple ways of imaging, you present many absent things together in one place, at the same time. Students scan through textual multimedia analyses, replay segments, pause in the midst of segments, and view slow motion segments. By making the "absent present" through slow motion, zooming, ghosting, and a variety of digital filters, we enhance our ability to focus our students' attention on underlying structures and patterns in the data. As Lynch has shown, "Representation includes methods for adding visual features which clarify, complete, extend and identify conformations latent in the incomplete state of the original specimen. . . . The object becomes more vivid; we can picture it as though it were 'naturally' present for our inspection" (1990, 181). One can enhance frames in the digital video, add arrows or outlines superimposed over moving animations, enhance the color distinctions of dyes and other discriminator conventions to help clarify and extend one's description and analysis of a process.

On the other hand, representation also includes methods for leaving out visual features and applying masks to parts of the image. For example, when a "find edges" filter is applied to a digital video clip, the result is a black and white sketch image of the scene highlighting with bold outlines the form and movement but completely masking most other identifiable markers. Or applying this filter to close-up footage of a plant growing would temporarily eliminate much visual information (color, texture) but would highlight the edges of the plant's structure as they grew and enlarged into new spatial areas. Therefore, in order to foreground an analytical point one wants to make, one can mask certain nonrelevant visual and aural data. Scott McCloud has noted that this kind of sketching or cartooning representation is a form of what he calls amplification through simplification: "When we abstract an image through cartooning, we're not so much eliminating details as we are focusing on specific details" (1993, 30).

I created abstract representations of points of analysis for my course's CD-ROM using digital filters and imaging software. In an instance when I had difficulty identifying a speaker out of a group, I used an "image-pan" filter to zoom in on the scene to observe whose mouth was moving. One could just as easily zoom in on the details of a cell wall beginning to divide. Throughout the CD, digital filters were used to (1) compile small collections of short examples of video segments edited together; (2) bring the action in closer to the viewer, although the original videotape was shot from a distance; (3) shift the camera angle; and (4) present identifying titles and some bits of transcription. And of course these "data clips" were presented alongside their textual analysis and any relevant still frames.

The educator can control and manipulate the visual data by, among other things, adjusting the speed of the digitized clip and slowing down or speeding up the action from 1 percent to 10,000 percent (contrasted with the limited range allowed on VCR toggle controls). In my courses this capacity has had a tremendous impact on the practice of repeat viewing and on observation and isolation of units of interaction, and this flexibility quickly becomes a capacity on which I routinely rely. Similarly, one can imagine being able to slow down or speed up an animated simulation of a steel beam under increasing stress; the pattern of increasingly weakening areas could be viewed repeatedly by students until they are able to recognize the pattern itself. In my courses some of the students also wanted to learn how to handle their own data digitally. They selected key segments from their videotapes, digitized them, and, among other things, created regular-speed and slow-motion movies like the ones they have examined on the CD-ROM.

HTML and Containment

Those readers conversant with the Internet might ask if the teaching and research practices that I claim are changed by CD-ROM technology might also be impacted similarly by hypermedia documents created for the Web. I considered creating an HTML document (a very large one) to post on the World Wide Web, but I sensed a potential problem with "containment." That is, since the document included hypertext links to other auxiliary pages, there was little assurance that anyone would ever read the study in a coherent and linear fashion. After all, with an argument to build and with evidence to present in an orderly and persuasive manner, one needed a medium that provided both the multimedia capacity required for showing animated examples and a contained medium that maintained the scholarly integrity and coherence of the study. I wanted to guide my reader along the path of the case I wanted to make, and the suasory force of my argument required that the supporting evidence be offered in a progressive sequence. The CD-ROM medium at least contains the content to my study, much as a book might, without risking the loss of a reader to a distant hyperlink and its links.

The size of my files was also a problem for the Web. Digital video and sound clips are large files, and even with highly optimized clips, some viewers would spend too long waiting for them to download in order to view them. CD-ROM technology was exactly the solution I needed; the file size problem disappeared entirely, given the 650-megabyte capacity of a CD, and I was able to "locate" and contain the whole study within the domain of one CD-ROM.

My argument for containment and control might be controversial in light of the counter notions of "freedom" and "nonlinearity" provided by hyperlinks in Internet documents. But consider the example of narrative: many readers ultimately do not want to have to invent a new ending or navigate through a maze of possible outcomes. Many readers simply want to be told a good story with a plot that builds on itself and unfolds in a logical way. The structure of such narratives, although carefully crafted, designed, and controlled by the author, is often most appreciated when this same structure remains largely transparent to the reader, when it does not get "in the way" of the unfolding narrative. Similarly, for teachers guiding learners through new content, new terminology, or new analysis, a logical and linear structure can be extremely helpful for the students. I suspect that after initial excitement over the freedom of nonlinearity dies down, many scholars, students, writers, and readers will hasten to return to a more linear, sequential, and contained structure. This is what CD-ROM technology provided me with for the dissertation.

Ideas Travel Fast: Rapid Access Near and Far

Rapid access to information is one of the key features of CD-ROM technology, and I cannot overstate the powerful impact it has had on the ways I practice my research and create presentations for the classroom and other venues. The impact of rapid access is very subtle and is often overlooked because once we begin to employ a technology that affords us rapid access to information, we tend to adapt to it and take it for granted very quickly. For example, as soon as one becomes acquainted with fast Ethernet connections when cruising the Internet, it can become viscerally frustrating to "return" to a slower modem connection. One may have been unaware of the degree to which one's own rapid adaptation to rapid access had actually occurred until that rapid access was no longer available.

In pursuing my immediate daily academic tasks—teaching today's class, comparing data sets—the fact of rapid access usually becomes transparent; I am not conscious of it. But in the moment-to-moment interactions with one's research materials and with students and other audiences, the capacity to have at one's command a host of information, files, video clips, and other data or examples tremendously enhances one's ability to conduct analysis and to teach.

For example, I have used CD-ROM technology as a tool in lecturing for my own classes, for seminars across campus, and for guest lectures at other universities. Hearing me read from a paper document complete with still-frame images from my dynamic data samples simply is not as compelling for my audiences as seeing and hearing the multimedia analysis and array of examples. In some instances, the host university has provided computer display technology that allows me to project the CD-ROM onto a larger screen while "walking through" its key features. This gives me great flexibility during question and answer sessions, because I can rapidly interact with the audience and the CD, clicking to various examples, jumping from one section of analysis to another, and responding in much more depth to spontaneous questions. For those readers who have used videotapes in presentations, it is worth recollecting how painful it can be to search through your analog videotapes to locate a "best example" in response to a question. This is no easy feat "on the fly." In fact, what we often do is give up trying to locate the clip we have in mind and simply try to tell our audience or students about it. The show-and-tell capacity of the CD is powerful in scholarly encounters as well as in the classroom.

Another often overlooked feature of CD-ROMs that my students and I have made use of is their utility as a storage medium for our digitized video data clips. Although this might seem trivial, in fact it has been an extremely important function and is directly related to the issue of rapid access. For students learning to manipulate their digital video data, to create collections of clips that have recurrent patterns of behavior on them, and to conduct micro-analysis of the behaviors, the CD-ROM has become a trusted way to store the most important clips while projects are in progress (and for future reference).

For example, using a CD-ROM as a digital storage bank for one's data clips, incidents, examples, or source cases, one can rapidly follow up on those glimmers of insight that are frequently abandoned otherwise. If the bit of data or the videotaped segment one is looking for is embedded somewhere in a two-hour analog tape, on some misplaced disc, or in a file drawer at home, one is simply less likely to stop one's current activity and make the physical effort required to locate the data set. If it is located on a CD-ROM, one can slip the CD in the loading bay, simply scan through the contents of folders to find the file-movie or data file that had come to mind in that insightful moment, and proceed with one's comparison or analytical activity. This truly rapid access to one's archive of information and knowledge should not be underestimated, since those glimmers of insight—if not acted on right away—tend to vanish into thin air. Being able to pull up a file for comparison and contrast with another case has enhanced the analytical momentum of my own research projects.

When preparing lectures for class, I have frequently loaded one of my storage CDs and scanned through the archived clips to find just the right example for class. The majority of key video data segments for my research have been digitized and stored on CD-ROM. Since this medium provides me with very rapid access to my digital video data files (as well as collections of text and image files), I can avoid the damage caused by repetitive viewing of the original analog videotaped materials. Once I locate the clip, I copy it for that day's class onto my hard drive from the CD, manipulate the clip as I wish for the class lecture, make a "new" movie just for this class, and "download" to videotape again the custom-designed "analysis movie" to use with the students that day.

So there are very practical advantages to using CD-ROM technology to store huge files in a way that facilitates rapid access. And as noted, the subtle but powerful fact of rapid access has an impact on the flexibility with which I can interact with the information on the CD, with students' and other scholars' questions, with my own glimmers of insight while conducting research, and with the creation of custom-designed presentations.

These three features of the interactive CD—its multimedia show-and-tell capability, its containment of content material, and its provision of rapid access capabilities on the fly—have improved my presentations, my teaching, and my research.

Dissemination

Part of our task as teachers-researchers is to disseminate knowledge. Interestingly, some of the more conventional features of CD-ROM technology have allowed me to send copies of my CD all over the world for other scholars, educators, and students to use. This is another kind of rapid access to information, but at a distance. Some fundamental features of this technology that have sometimes been left out of discussions of functionality include the extremely low cost of CD duplication as contrasted with paper duplication, its simple packaging, and its small size and minimal weight for shipping concerns. My CD, which contains just over 400 megabytes of text, still images, and video clips, is being examined and used in classrooms in the United States, the Netherlands, the United Kingdom, Japan, Sweden, Ireland, Germany, India, Spain, Canada, Indonesia, Australia, France, Brazil, and Denmark. Its basic duplication costs were approximately five dollars per unit when I created a batch of 250 copies. This would simply not have been possible with print technology. The interactive capacity for rapid access to information on the CD-ROM and the easy replicability have allowed scholars to disseminate research to wider audiences, near and far, more cheaply and quickly.

In this essay I have described some of my experiences using CD-ROMs in research and in the classroom. In particular, I have shown some concrete ways that this technology has changed two academic practices and what properties of CD-ROMs have been most important to these changes. This technology has enhanced the ways I present instructional material in class and at scholarly meetings because I can provide students (and other scholars) with powerful new representations of the phenomena about which I teach alongside my analysis of the phenomena. Whether digital video clips or computer-generated simulations, these new representations give students greater access to the dynamic and sequential aspects of whatever process-oriented subject they might study. Furthermore, the boundaries of the classroom are extended in that the student can interact with the CD-ROM away from class, individually, and can probe to greater depths its content. My students have also demonstrated through their questions and presentations that analytical and methodological lessons have been taken up from the CD-ROM as well.

Finally, many educators spend a great deal of time collecting, storing, organizing, manipulating, reproducing, and disseminating various kinds of data. The study of any process that involves dynamic components lends itself to fuller examination through a CD-ROM's ability to bring together in one "space" multiple representations; the representations can be aural, visual, sequential, textual (inscribed), and dynamic in nature. Depending on one's immediate purpose, the teacher-researcher can enhance what is shown or mask and leave out distracting parts of a representation. CD-ROM technology offers a powerful new way to array one's evidence, arguments, and examples; it involves showing and telling, together in one location, and making the "absent present." The technology also allows one to have rapid access to digital data of all kinds and to disseminate one's work cheaply and efficiently in an enhanced format. I am still exploring and learning about its possibilities, and my own way of teaching and conducting research will never be the same.

WORKS CITED

Jarmon, Leslie H. 1996. An Ecology of Embodied Interaction: Turn-Taking and Interactional Syntax in Face-to-Face Encounters. Ph.D. diss., on CD-ROM, University of Texas at Austin.

Latour, Bruno. 1990. Drawing Things Together. In *Representation in Scientific Practice*, ed. M. Lynch and S. Woolgar, 19–68. Cambridge: MIT Press.

———. 1986. Visualization and Cognition: Thinking with Eyes and Hands. *Knowledge and Society* 6: 1–40.

Lynch, Michael. 1990. The Externalized Retina: Selection and Mathematization in the Visual Documentation of Objects in the Life Sciences. In *Representation in Scientific Practice*, ed. M. Lynch and S. Woolgar, 153–86. Cambridge: MIT Press.

McCloud, Scott. 1993. *Understanding Comics: The Invisible Art*. Northampton, MA: Kitchen Sink Press.

Polanyi, Michael. 1967. *The Tacit Dimension*. Garden City, NY: Anchor Books.

Chapter Eleven

Doing Theory in Hypermedia Practice

A Case Study of the HyperHistory Video Project

Lisa Cartwright

In their 1996 collection of essays about contemporary video, Michael Renov and Erika Suderberg remind us that emergent technologies are never as new as they appear to be; they are experienced in relation to older and more familiar media, which they challenge and destabilize. Independent video, they suggest, must be studied in light of its challenge to the medium's historical role as a tool of surveillance and a key technology in the ascendance of a one-way communications paradigm (xii). I want to bring this issue to bear on the discussion of CD-ROMs and hypermedia in the media studies classroom. In this last decade of the century we've seen great interest in the use of media technologies for learning and teaching. Far from new, this interest can be traced back to experiments with video in networked classrooms and specialized programming for schools during the 1950s and 1960s (Goldfarb 1998, chaps. 1 and 2). Like the disparate projects that make up the currently burgeoning instructional technologies market, these earlier efforts ranged from conservative attempts to standardize curricula and increase profits through distance learning, to progressive initiatives to address the educational needs of specific constituencies and enhance the teaching-learning environment.

Aware of this history, I witnessed with tempered enthusiasm the emergence in the early 1990s of CD-ROMs and other new media forms as curricular tools in higher education. Although many of the programs available for teaching were designed to promote interactivity and participatory learning, for the most part this participation was limited to selecting options for movement within a narrative or a set of data—hardly a major step toward including users in the production of knowledge. My impulse was to look for ways to destabilize and challenge this limited idea of interactivity and the techniques of the emergent instructional technologies industry generally, following the model of what independent video producers had done in relation to television from the 1960s through the 1980s. Rather than limiting themselves to the conventional format of written analysis in their critical work, intellectuals like Martha Rosler, Laura Kipnis, and Dara Birnbaum chose to use video production to contribute to scholarly/activist challenges to mainstream media culture. How, I wondered, could I incorporate into my course this strategy of intervention in the means of production without falling back on the use of media texts (whether alternative or mainstream) solely as objects of analysis? Could students be trained in an academic, nonproduction course to produce scholarly texts in a medium other than the conventional paper—specifically, in some form of hypermedia—and could this work bring something new to our ways of thinking about how we regard the media we study and teach?

In the fall of 1994 I offered a graduate seminar at the University of Rochester on video art and politics; this course became a venue for addressing these questions. The course was designed to consider the history of alternative video, much of which had not been written into the film and television studies canon. The work I had in mind included video by collectives such as Ant Farm, Paper Tiger Television, Testing the Limits, and the Inuit Broadcasting Corporation. One of my goals was to create a critical dialogue around relatively underconsidered aspects of video history, and to historicize media projects that have operated in the margins of the mainstream art and television markets. This course would consider the intersecting histories of guerrilla television, Portapak activism, and the more thoroughly canonized area of art video, as each of these areas informs the current status of media innovation and activism.

Organizing such a course in the early 1990s was not easy. Graduate-level readings were hard to find. While more recently some outstanding volumes on alternative video have been published (Renov and Suderberg 1996; Boyle 1990), a canonical body of texts could hardly be said to have existed just a few years ago. There were numerous art exhibition catalogs and monographs and a few critical-theory essays that had been granted canonical status in art history—texts like Rosalind Krauss's 1976 "Video: The Aesthetics of Narcissism" or Martha Rosler's 1985 "Video: Shedding the Utopian Moment," or the numerous catalog essays and journal articles by David Ross. However, there were very few substantial video history and theory books to draw on.[1]

As a whole, this body of work did not add up to the kind of intensive materials one would expect to find in a graduate course. My strategy, then, was to shift the emphasis from textual consumption to hypertextual production. One of the framing issues of the class was discussion about the historical conditions that had prevented a comprehensive theory and history of video to emerge. I was teaching a number of students who had worked in or written about independent video, public art, and media activism, and so I hoped to draw on their experience and research to supplement the available texts.[2] My approach was to get the class to think about generating, for themselves and others, exactly the kind of work we found missing in published texts—in other words, to use the course to build new materials for the field.

This approach immediately raised doubts for me about the adequacy of the conventional class paper to the goal of training graduate students. How might papers about obscure video be received by academic, arts, and cultural studies journals—a context where there was a lack of "authoritative texts" to ground discussions about alternative video? What field, or what set of debates, would this work be in dialogue with? These problems were augmented by the fact that many of the videos we planned to study either were narrowly distributed (as in the case of narrowcast programming for Iranian cable shows) or were not preserved for distribution—a situation that is steadily changing thanks to the work of key individuals and archives.[3] Even when I was able to track down tapes, we were faced with the problem of writing about video that had limited audiences even at peak exhibition. Hence writing about this work had to be done in conjunction with making the videos themselves accessible and more relevant to a more general film and video readership.

Hypermedia projects seemed to make more sense than papers because they would allow the projects to incorporate images from the videos they discussed, making it possible for viewers to see at least some of the work. I also (mistakenly) believed that, unlike a book, a hypermedia project could circulate almost immediately upon its completion (in the end we encountered copyright problems that have prevented it from circulating at all). Most important, though, this kind of project would allow some students to bring their production experience to bear on the making of historical or theoretical texts, giving them a means to work through a core formal problematic of alternative media practice. Thus the collective production of a CD-ROM on aspects of independent video history seemed like a solution to a number of the problems I encountered in organizing the course.

This project also raised some more general questions about pedagogy: How do we teach media works such as alternative videos that fall outside the canon of an already canonically marginal field like film studies? What constitutes "legitimate" scholarly graduate student work in an English department course on video? Hypermedia has made its way into the film and television studies classroom as a related media form to be analyzed using the critical vocabularies of film, television, or literary studies, or as a pedagogical resource for presenting films and analytical writings together (e.g., Rick Preilinger's fascinating CD-ROM culled from his archive of industrial and educational films, or *The Rebecca Project*, a wonderful interactive CD-ROM by Lauren Rabinovitz and Gregg Easley also published by Voyager). But it is not yet clear whether hypermedia can become an accepted way for students themselves to produce texts, much less to make work that will be accepted as evidence of their professional development as media theorists or historians, rather than as proof of their allegiance to the "wrong" side of the theory-practice divide. These questions are reminiscent of 1970s-80s discussions about the scholarly and political work of "theory films" (the work of Laura Mulvey and Peter Wollen, for example) and the promise and limitations of doing history and theory in audiovisual media. An impediment to the continuation of this discussion is the historical institutional divide between media production departments and history/theory programs—a split that is familiar to most of us who teach film and media studies in universities, and which I won't belabor here. But for those of us interested in questioning the form our scholarship takes, or pushing the envelope of scholarly activities into realms including public and alternative cultural spheres, it is essential to begin to rethink what should constitute a scholarly text, and what forms, including CD-ROMs and Web pages, might challenge certain paradigms of the field, or make relevant the issues we raise in our essays to communities beyond the academy. My discussion below about the work my class produced is intended to address, if not answer, some of these questions.

Before turning to a discussion of the HyperCard projects themselves, however, I would like to make two brief and related points. First is the rather obvious detail that access to the means of production is no small consideration in envisioning how students and teachers of media studies might use hypermedia production to begin to address some of the issues raised above. My course was offered at a moment in my (medium-sized, private) university's institutional history when few humanities courses were taught with a hypermedia component other than classroom multimedia displays or study guides. At the time we had Mac and IBM classrooms but no media lab for independent student work. Members of our class had access to two Quadras, a video deck, and a scanner—all reserved for faculty use and made available to students by special arrangement. The university purchased a CD writer one year after our project was completed. The two computers on which we produced and stored our work did not have the memory to store real-time images for this many projects. It was therefore by necessity that we agreed to limit our projects to the relatively low-tech format of HyperCard, using color, graphics, and sound but no moving images. In the years since this course was taught, we've upgraded considerably; the kind of projects we produced would now be a piece of cake for students, many of whom are quite skilled in authoring for the Web. Indeed, the rise of the Web as a venue for making data and creative work public has posed a serious challenge to the CD-ROM, which we originally envisioned as the final form of our project. That this medium may go the way of the eight-track audio tape is evidenced in the current reluctance of publishers to take on CD projects, and the recent cutbacks at the major CD developer and publisher Voyager.[4] Nonetheless, I believe that the overall project, despite its failure to circulate as widely as we had hoped, still stands as an example of the kind of innovation in student work that can occur at the low-tech end of the scale when hypermedia is brought into the curriculum as a tool for independent research and intellectual production, rather than solely as an object of analysis or a classroom lecture illustration. The real value of the project was in the process through which students experimented with scholarly form and the limits and possibilities of hypermedia composition.

My second point concerns the status of video history and hypermedia in the canon. Whereas video is perhaps the most undertheorized medium in film and television scholarship, in its relatively short existence hypermedia—in both its utopian and its dystopian potential—has been the subject of a virtually instant canon during the 1990s. At this writing, I would wager that there are already more books and journal essays devoted to computer media than to the much longer history of alternative video.[5] A few authors, most notably Lynn Hershman Leeson, present the two forms in relationship to one another;[6] however, it is much more common to see hypermedia described as either a literary tradition, with its roots in the technological history of the printing press,[7] or a photographic tradition, with its roots in the still camera.[8] A third approach, with ties to the field of communication and the historical critique of the rise of the television industry, poses a kind of latter-day Frankfurt school critique of computer-based media, citing the advancement of global capitalism via new technologies. While all these approaches are valid and useful, we can learn much about the history and potential of hypermedia, particularly as a combined text-and-image medium of political critique, from the example of alternative video. It is worth noting, at the very least, that video prefigures computer hypermedia insofar as it includes text, image, and sound in one medium that can be manipulated (and not simply viewed) by the individual or home consumer. Moreover, hypermedia can be said to incorporate video insofar as computers increasingly are equipped to perform the functions of recording and playing back sound and moving images in addition to the more conventional task of static text and image generation and display. To return to a variation on Renov's and Suderberg's point, we have much to learn by considering hypermedia in the historical framework of alternative video.

The Class

My twenty-student graduate seminar, titled Video Art and Politics, was presented with the option of producing thirteen-page image/text works that could be adapted to HyperCard, a program that was available at our university and had a relatively less steep learning curve than, say, Macromedia Director. These projects, I suggested, might serve as introductions to a particular area of video or media history, or they might provide a look at an underconsidered moment of video or multimedia history and politics. Less than half the class followed through on this option. This was not a problem, since we did not have the equipment to support more projects, but the reasons for this limited interest are worth recounting. Technical instruction was not a part of the seminar. Students had to be able to make the commitment to learn HyperCard on their own time. Some students felt that learning to program and design hypermedia would do little to advance their profile as scholars—a problem that would be compounded by the fact that a project amounting to thirteen pages of written text would not allow them to develop their ideas as fully as they might in a conventional publication-length paper. We spent a lot of time discussing the goals of writing a more concise text, and what one could reasonably expect to accomplish in this format. Audience was also an issue we addressed at length: who would see and use these HyperCard projects? I will say more about these points in my discussion of the HyperCard stacks themselves.

The seven people in the class who pursued the hypermedia option formed a production group that worked in close collaboration for three months. Ironically, the project benefited from the low-tech facilities and its position in the curricular and canonical margins. This group initially met with Brian Goldfarb, who was teaching computer art as a graduate instructor, to learn how to compose HyperCard stacks. In some cases members of the group worked individually with Goldfarb to learn the technology, then produced their own stacks. In other cases Goldfarb collaborated with group members, who provided text and image sources and a rough idea of how they wanted their stack to work. Overall, the group worked collaboratively; members pitched in or provided instruction or feedback at different phases of each project. The videos and hypermedia projects we screened for the class became something more than objects of critical analysis; we began to analyze these works' ability to convey history, theory, and ideas more typically covered in conventional written forms. These examples presented some of the possibilities and limitations of working in an electronic medium with limited facilities and funds. In some ways, the principles of collective production modeled in much of the alternative video work we had considered were adopted by the group, which half jokingly took on an official collective identity, the HyperHistory Video Group. Working together in a computer lab fostered an atmosphere of interactive exchange of skills and ideas, and allowed for spontaneous group critique. I am sure this kind of spontaneous exchange would not have happened if we had had access to the media lab and home computers that are available to students now. The situation contrasts strikingly with the history of consumer home video. Whereas consumer video has remained a "home movie" phenomenon, right down to a lack of interest in home editing equipment and public exhibition, consumer hypermedia production has become a much more public and sophisticated affair—note the level of editing and design that goes into your average "family" Web page, for example. We maintained an aspect of the artisanal in our projects that was more tied to video's early history than to the digital culture we were a part of.

The Stacks

The seven completed programs, which we grouped under the title of the HyperHistory Video Project, range in topic from studies of underconsidered aspects of video history to critiques of narrowcast or public television programming. In terms of technique, they range from more conventional exercises in assembling data to experiments with the organization of theoretical and historical knowledge via sound, text, and image. I will describe a few of the projects below with an eye to demonstrating how students addressed, in different ways, some of the questions about scholarship, pedagogy, and video or hypermedia history raised in this essay's introduction.

One of the goals of the project was to generate material about underconsidered aspects of video history. Ed Chan, an English graduate student, took up this agenda in his production of a HyperCard stack about the history of guerrilla television. Drawing on primary documents (manifestoes, production stills, and historical video footage) and historical essays on the topic, Chan interrogates the notion that video history can be split into two tendencies—art video and alternative television. He reconsiders the goals and practices of early guerrilla and alternative TV, demonstrating continuities across genres of video practice. In its design, Chan's stack is the most straightforward of the set. Following a conventional branching structure, the stack allows users to select from a series of headings and subheadings to access ever more specific, primarily text-based information and essay-form arguments about the topic. The decision to work in a fairly simple, conventional format was a practical one: Chan taught himself HyperCard and programmed the stack on his own in a short period of time; hence his project was a technical learning exercise as much as it was a research project. Chan's project is remarkably strong as an example of a first, self-taught hypermedia project.

I noted earlier that one goal of the overall project was to make underconsidered, narrowcast, or community-specific media accessible and relevant to a broader readership. Ondine Chavoya, who now teaches art history at the School of the Museum of Fine Arts in Boston, created an image and text history of Asco, an underconsidered Chicano/a art collective whose important work of the 1970s was not brought to the broader attention of media and art scholars until the late 1990s, in part through Chavoya's own scholarship. Chavoya's task was to make apparent Asco's relevance to media history while also making accessible to viewers the group's little-known history, texts, and images. The stack Chavoya assembled with Goldfarb focuses specifically on the collective's No-Movies, a series of performances and photographs documenting the production of "movies" that the collective never actually made. As Chavoya explains, No-Movies were both a parodic send-up of the Hollywood cinema, produced a few miles from Asco's community of East Los Angeles, and a wry commentary on the exclusion of Chicano/as from the mainstream industry and from the means of making films. One of the successes of Chavoya's project is its clear demonstration of how an account of the No-Movies belongs in media history precisely because of the media forms the collective pointedly did *not* use.

Babak Elahi assembled materials for a stack about narrowcast Iranian television, borrowing from Hamid Naficy's recent book (1993) on the topic. Like Chavoya's, Elahi's stack required assemblage of hard-to-access visual materials, and aimed to reproduce media for an audience that might not otherwise have access to the work. Elahi was able to get tapes of the shows and permissions from their producers, allowing for fairly extensive use of frame enlargements throughout the stack. Ultimately, Elahi intended the stack he produced with Goldfarb to be a teaching accompaniment to Naficy's book. Readers who were not familiar with the programs Naficy described could access Elahi's stack and see images of the shows and their performers while also gaining Elahi's insights on a facet of this work considered more briefly in Naficy's book: family values as they are reproduced in U.S. Iranian cable television shows. Elahi included response screens in the program, where users could write their thoughts on the point being made or view the responses left by previous users. With this precise goal of providing support for a book and fostering teaching of the topic, Elahi's stack began to move away from the conventional textual analysis with frame enlargements to a more directly pedagogical orientation. As he wrote the text for the stack, Elahi struggled with the desire to make a more complex analysis of the compelling material, as well as with the limitations imposed by the form. Like others in the group, Elahi found it hard to break away from the model of a text-heavy program in which images are present in order to be subject to analyzed.

Elahi's predicament recalls some of the questions posed earlier in this essay: Could hypermedia bring something new to our ways of thinking about how we regard the media we study and teach; and could we resist falling back on the model of using media texts solely as objects of analysis? Tina Takemoto's stack, titled "Double Eyelid: Cultures in Surgery/Cultural Insurgency," engages with these questions directly. A performance, mixed media, and installation artist who is also an art historian and critic, Takemoto is invested in finding ways to use critical writing, visual art, and performance techniques together to address the problematic divide between studio practice and art history/theory. The stack was a means of finding ways to "do theory" in visual practice or to produce historical and theoretical texts in something other than the conventional written format. Using the visual as material to subject to a written textual analysis is precisely what Takemoto did not want to do. Her goal was to use the graphics and multimedia capabilities of the computer to bring something new to the process of critical analysis. Because this stack addressed so successfully many of the questions that concerned me in devising this project, I will consider at greater length what Takemoto accomplished, as well as the logistical problems she and others encountered.

The stack addresses the larger issue of stereotypical representations of Asian femininity in Western media cultures by interrogating the cultural politics of elective cosmetic surgery undergone by some Asian and Asian American women to make their eyes appear "more Western." The stack attempts to get beyond the thesis that Asian women who opt for eyelid surgery are simply victims of Western beauty culture, making a case for a more complex understanding of the issues involved. Although it presents a group of videos on the topic, "Double Eyelid" also stands on its own as critical analysis and work of hypermedia art. Takemoto appropriated text and images from medical textbooks and videos about eyelid surgery and cosmetic surgery generally, selecting salient passages that reveal particularly problematic ideas about race, physiognomy, and beauty, and intercutting or juxtaposing these segments with graphic word plays and quotes from a critical essay on the subject by the theorist Kathleen Zane. Frame enlargements and sound clips from videos by Valerie Soe, Pam Tom, and Tran T. Trang comprise a section of the stack that functions as a kind of annotated videography. Takemoto uses subtle graphic word play, composition, and editing of sound, text, and image to suggest a critique of the practice without spelling out that critical analysis in so many words for the program's users. The stack is a unique example of the way editing, composition, and plays with juxtaposition and quotation of sound, image, and graphic elements can convey a critical analysis without extensive support from a spoken or written "theory text."

In addition to her goal of producing a critical analysis fully using the formal possibilities of the medium, Takemoto wanted to make available information about a small group of relatively underconsidered videos—those devoted to an interrogation of race and gender issues in cosmetic surgery and beauty culture. Reception of the portion of the stack devoted to artist videos, however, was complicated. Whereas the use of found medical books and videos relies on the familiar alternative video strategy of found-materials appropriation, the section devoted to presenting the work of Soe, Tom, and Trang was not meant to be read as textual poaching. What was intended as promotion of artists' work, however, was received by at least one producer as another artist's appropriation of her work (or so we learned informally, through bicoastal gossip). This suggests that the reception of Takemoto's stack was complicated by her mixing of both professional identities and textual strategies: she made the stack as both historian/critic and artist, and as such it crosses the genres of scholarly analysis and art. Because it combines approaches familiar to video production (attention to visual composition and design, the use of appropriated footage) with those of the media historian and critic (the presentation and analysis of archival materials), the stack can be read as either, or both, video history/criticism and a work of media art. To read the project within the new mixed genre of hypermedia critical history/visual art practice requires an audience of users with an understanding of the crossing of disciplines and identities that goes into this sort of project.

The predicament of intentions versus reception illustrated by my account of Takemoto's stack is indicative of a more pervasive problem faced by hypermedia producers regarding the fine line between reproduction toward the goal of promotion and out-and-out appropriation. Clearly there are legal issues at stake here—issues that we as a group did not adequately address in our production, a fact that left us unable to pursue publishing venues for the project. Rather than address the important question of image copyright, however, I prefer to take up the rather underconsidered issue of how we (academic users of hypermedia) tend to classify the images we reproduce in our work. Images reproduced in the conventional format of the illustrated written essay could hardly be said to constitute creative appropriation of the work of an artist, as long as rights are secured and fees are paid. The work typically serves the purpose of illustration or object of critical analysis. What happens to this same work, however, when it is incorporated into a hypermedia text that works with precisely the issues of context, juxtaposition, and framing to make its point? Does incorporation of the work into the body of the critical analysis compromise the status of the work? Would such use constitute appropriation in the sense that the integral meaning of the work is compromised? In the HyperHistory Video Project, one stack, produced by Alison Schultheis, allows the user to manipulate video clips from a copyrighted PBS documentary. The clips can be zoomed in on and otherwise altered in a manner that most definitely changes the structure of the original program. In this case, the analysis of the program unquestionably results in various alterings of the original work, resulting in variations as diverse as the number of users who encounter the program. This level of user interaction with the text under analysis opens up potential problems beyond the simple one of reproduction rights. As the case of "Double Eyelid" suggests, these problems are not just legal but ethical, insofar as media historians and critics have an obligation to respect the integrity and autonomy of the producers and works they write about.

There is more to be learned from the project's problems in terms of questions of audience. The HyperHistory Video Project as a whole tended to fall between the cracks of the alternative video community, to whom the project's content was pitched, and those invested in hypermedia culture, who were less likely to be interested in the issues of low-tech, community-based programming addressed in many of the stacks. This question of reception was complicated by the fact that the people involved in promoting alternative video history and community-based video—particularly those who have embraced a Marxist critique of technology—have not been the first to embrace the high-production-values aesthetic and pro-technology sensibility that dominate in hypermedia culture. To step back and read this stack as a media critic, I would argue that it successfully demonstrates the potential to "do theory" in a format that combines sound, the visual, the graphic, and the textual, suggesting that there are ways for media historians and theorists to use hypermedia to go beyond the conventions of the scholarly essay. However, if this possibility is to evolve, we need to address the questions of which communities will respond to this medium, how readers/viewers will receive these texts, and what sorts of genres and conventions will become accepted (among media historians generally, not just within the academy) as hypermedia takes hold more pervasively.

I'll conclude by returning to some of the more strictly pedagogical questions I began with. In the graduate programs in which I teach (English and visual and cultural studies), we have yet to see our first hypermedia dissertation. This is understandable; for such a project to come out of the humanities would require a formal revision of our university's dissertation standards; moreover, students can hardly be expected to take the risk of experimentation with a project on which their future livelihood depends. In a job market where promotion is dependent on publications, the lack of publishing venues for hypermedia makes the form a poor choice for a dissertation. To see this situation change over the next few years, we would need to address, in addition to some of the questions raised immediately above, the issues posed earlier regarding the ability or willingness of scholarly publishing venues (presses, journals) to support hypermedia. At the moment, the possibility of change on this front among publishers seems less than promising, due to prohibitive production costs, the likelihood of weak returns (due to the fact that hypermedia is easily reproduced), and strong resistance among academics themselves. Ironically, while publishers are turning away from the marketing of items like CD-ROMs, there has been a move to make dissertations available on the Web. This situation has raised an outcry among those with the biggest stake in maintaining rights to knowledge (scientists rightfully fearing corporate raids on their new research, for example).[9] It would be easy to chalk up to conservatism faculty resistance to the turn to hypermedia in the classroom. But this is not just a matter of the old boys sticking their heads in the sand. Some of the most progressive among us also recognize the threats of deskilling, enforced retraining, and work escalation posed by the promise of new educational media to expand revenues for universities.

These issues notwithstanding, I've tried to show that there is much to be said in favor of the excitement of venturing into the creation and critique of visual and graphic forms in conjunction with written text, and the break with regarding the image solely as object of analysis in the media studies classroom. And too, perhaps hypermedia will take us beyond the academy's historical disdain of media (specifically television) as low culture. But it is important to note that one needs to invest a lot of personal time, effort, and resources to learn new technologies and to pull off successful hypermedia projects—particularly when the production training and advising of individual students are part of the picture. We invested a huge amount of night and vacation time and personal resources in the HyperHistory Video Project, a fact that makes me less than eager to reproduce the experience without significant increases in institutional support for faculty- and student-produced hypermedia. Perhaps one indicator of the direction of things to come is that while the ratio of Web-based projects to papers has sharply risen in my end-of-semester undergraduate coursework pile, I find fewer media-focused projects that express concern regarding access to the means of production and the ways that knowledge changes in relation to changes in resources and the textual composition process. Graduate students' hypermedia work for the most part has been limited to designing university Web sites for a fee or developing classroom displays—projects that delimit hypermedia production in the realm of professional service (the innovative side of classroom display production notwithstanding). This is troubling to me in part because I teach in a graduate program dedicated to the study of the visual in cultural studies; what better site from which to pose the question of the place of digital media in scholarship? Perhaps it is inevitable that excitement in hypermedia is bound to fade as instructional technology use becomes part of the university's official development plan; for whatever reason, I and my students are less focused on experimentation, formal innovation, and the historical circumstances against which the medium appeared so compellingly different and promising. The bottom line is that a challenge to the ideological orientations of the new technologies entering the educational marketplace requires a paradoxical combination of circumstances that is difficult to find or sustain. On the one hand, we need institutional support; on the other hand, we need autonomy from the larger agendas of the institutions we work for, and the larger instructional media marketplace generally.

NOTES

1. The one comprehensive anthology of writings by video theorists, historians, and producers that we used was Hall and Fifer's *Illuminating Video* (1990). The bulk of my syllabus was made up of essays culled from catalogs and art magazines, specialized anthologies like Thede and Ambrosi (1991), classical collections like Hanhardt (1986), and a couple of theory books like Cubitt (1991).

2. For example, Laura Marks, who was a part of this group but did not complete a hypertext project, had done some innovative writing on the Inuit Broadcasting Corporation, among dozens of other alternative video producers. See her book on experimental film and video (forthcoming from Duke University Press).

3. Electronic Arts Intermix, among other institutions, has a program to preserve historically important video. Individual efforts to preserve neglected work include media historian Chon Noriega's work on the preservation of the video of Chicano artist Harry Gamboa, Jr.

4. Erkki Huhtamo (1996) argues the opposite point. He writes that the CD-ROM has the best chance among new media forms of making an impact on the consumer market. He does not, however, provide evidence to back up this argument—understandably, since his essay is devoted to the CD-ROM as an art form.

5. Again, Huhtamo makes the opposite point. In "Digitalian Treasures" he argues that "where such technological genres as video art have attained an

established role in the art world . . . the position of cyberartworks is often precarious" (1996, 306). While this may be an accurate reflection of the relative state of these forms in the art world until about the mid-1990s, we have to take into account the fact that video predates the CD-ROM; enough time has lapsed for video to become ensconced in the fine art tradition. By the second half of the 1990s, however, digital technologies had moved to the locus of major museum funding initiatives, occupied dedicated exhibition spaces, and been the focus of major exhibitions. Moreover, whereas video was never a medium that significantly served corporate interests, digital technologies are playing a major part in the global industries that subsidize the art world. These corporations' interest in the medium will likely ensure that digital art will maintain a special, if not central, place in the art world—a position that the far less ubiquitous medium of video could never have held.

6. See Leeson, "The Fantasy beyond Control" (1990). She notes that "a precondition of video is that *it does not talk back*," a fact that led her to "search for an interactive video fantasy" (267) that, perhaps, is better suited to the newer technologies about which she now writes. The contributions of Nell Tenhaaf and David Tafler to Simon Penny's excellent edited volume (1995) also bring together video and hypermedia history in interesting ways.

7. See, for example, Richard Lanham, *The Electronic Word* (1993), a book that includes a chapter on the visual arts but ultimately centers hypermedia history in the paradigm of reading and the legacy—and fate—of books.

8. For example, William J. Mitchell (1992) places the digital squarely within the classical history of photography, beginning with Talbot and the camera obscura. *Electronic Culture: Technology and Visual Representation*, edited by Timothy Drucker (1996), merges this gap between the written and the visual despite the bias toward the visual implied in the book's title.

9. On this issue, see *Educom Review*.

WORKS CITED

Boyle, Deidre. 1990. A Brief History of American Documentary History. In Hall and Fifer, 51–70.

Cubitt, Sean. 1991. *Timeshift: On Video Culture*. London: Routledge.

Drucker, Timothy, ed. 1996. *Electronic Culture: Technology and Visual Representation*. New York: Aperture Foundation.

Goldfarb, Brian. 1998. Media Pedagogy: New Technologies and Visual Culture in Postwar Education. Ph.D. diss., University of Rochester.

Hall, Doug, and Sally Jo Fifer, eds. 1990. *Illuminating Video*. New York: Aperture Foundation.

Hanhardt, John, ed. 1986. *Video Culture: A Critical Investigation*. Rochester, NY: Visual Studies Workshop/Peregrine Books.

Huhtamo, Erkki. 1996. Digitalian Treasures, or Glimpses of Art on the CD-ROM Frontier. In *Clicking In: Hot Links to a Digital Culture*, ed. Lynn Hershman Leeson, 306–17. Seattle: Bay Press.

Krauss, Rosalind. 1976. Video: The Aesthetics of Narcissism. *October* 1.

Lanham, Richard. 1993. *The Electronic Word*. Chicago: University of Chicago Press.

Leeson, Lynn Hershman. 1990. The Fantasy beyond Control. In Hall and Fifer, 267–73.

Marks, Laura. Forthcoming. *The Skin of Film: Intercultural Cinema, Embodiment, and the Senses*. Durham, NC: Duke University Press.

Mitchell, William J. 1992. *The Reconfigured Eye: Visual Truth in the Post-Photographic Era*. Cambridge: MIT Press.

Naficy, Hamid. 1993. *The Making of Exile Cultures: Iranian Television in Los Angeles*. Minneapolis: University of Minnesota Press.

Penny, Simon, ed. 1995. *Critical Issues in Electronic Media*. Albany: SUNY Press.

Renov, Michael, and Erika Suderberg, eds. 1996. *Resolutions: Contemporary Video Practice*. Minneapolis: University of Minnesota Press.

Rosler, Martha. 1985. Video: Shedding the Utopian Moment. *Block* 11.

Thede, Nancy, and Alan Ambrosi. 1991. *Video the Changing World*. Montreal: Black Rose Books.

Contributors

Scott Bukatman is an assistant professor of art history and comparative literature at Stanford University, and is the author of *Terminal Identity: The Virtual Subject in Postmodern Science Fiction* and the British Film Institute volume on *Blade Runner.* His essays (on science fiction, Jerry Lewis, superheroes, and the like) have appeared in *October, Camera Obscura, IRIS, Architecture New York,* and numerous anthologies. He claims to have taught Brian Kelly everything he knows.

Lisa Cartwright is an associate professor of English and visual and cultural studies at the University of Rochester, where she teaches courses in film, media, and science and technology studies. She is the author of *Screening the Body: Tracing Medicine's Visual Culture.* Her essays on medical imaging and communications have appeared in *Representations, Zone, Camera Obscura,* and elsewhere. She also produces educational computer programs on health and media topics.

Ted Friedman is a doctoral candidate at Duke University's Literature Program and a freelance cultural critic. Selections from his work can be found at his home page, http://www.duke.edu/~tlove.

Vanessa Gack is a doctoral candidate in the department of communication at the University of California in San Diego. She developed her micro-analytic approach to the study of kids' after-school interactions around computer games with help from the Mellon Foundation, The Institute for Research on Learning, and the Laboratory for Comparative Human Cognition. She is currently writing her dissertation on that topic.

Leslie Jarmon is a special lecturer for the graduate school of the University of Texas at Austin and a research associate at Indiana University at Bloomington at the Research Center on Language and Semiotic Studies. She received her Ph.D. in speech communication from the University of Texas at Austin. Her dissertation is the first in the United States to be published on CD-ROM, and her growing use of digital video technology influences both her research and teaching. Jarmon's interest in human communication grew out of undergraduate study in theatre and eight years in the U.S. Peace Corps. She is also a National Service Fellow research grantee.

Henry Jenkins is the director of the Comparative Media Studies Program at the Massachusetts Institute of Technology. He has published two ethnographic studies of media fan communities, *Textual Poachers: Television Fans and Participatory Culture* and *Science Fiction Audiences: Watching Doctor Who and Star Trek*. He is the editor of *The Children's Culture Reader*, also available from New York University Press.

Brian Kelly is a doctoral candidate in American studies at the University of New Mexico and received his B.S. in mechanical and industrial engineering from the University of Illinois. His work deals with technology and lived experience. He claims to know a few things that Scott Bukatman did not teach him.

Janet Murray is a senior research scientist and director of the Laboratory for Advanced Technology in the Humanities at MIT. She is an award-winning multimedia developer, and has published widely on both humanities computing and Victorian studies. She holds a Ph.D. in English from Harvard University. She has taught interactive fiction writing at MIT since 1992 and is the author of *Hamlet on the Holodeck*, about the aesthetics of the emerging digital narrative art.

Angela Ndalianis lectures in cinema studies in the School of Fine Arts, Classics, and Archaeology at the University of Melbourne. She is completing her Ph.D. on contemporary Hollywood entertainment structures, particularly exploring the blockbuster special effects cinema and the interrelationship among films, comics, and computer games.

Greg M. Smith is an assistant professor of communication studies at Carlow College in Pittsburgh. He received his Ph.D. in communication arts from the University of Wisconsin-Madison. His work has appeared in *Cinema Journal, Journal of Film and Video,* and elsewhere, and he is coediting (with Carl Plantinga) *Passionate Views: Film, Cognition, and Emotion.* Before entering academic media studies, he worked for several years as a software engineer for IBM.

Alison Trope is a doctoral candidate in the School of Cinema-Television at the University of Southern California. She is writing her dissertation on museums and film.

Pamela Wilson is an assistant professor of communications at Robert Morris College in Pittsburgh. She brings a background in cultural anthropology, folklore, and cultural history to her current work in the cultural studies of media. With a Ph.D. in communication arts from the University of Wisconsin-Madison, she has a special interest in the cultural politics of nonfiction media, especially as they pertain to indigenous, ethnic, and regional subcultures. She has gained a newfound respect for amateur historians through her work with online genealogical communities. Her work has appeared in *Camera Obscura, South Atlantic Quarterly, Quarterly Review of Film and Video,* and elsewhere.

Index

Lightning Source UK Ltd.
Milton Keynes UK
UKHW011955100320
360079UK00016B/295